BIOMARKERS

AND OCCUPATIONAL HEALTH

PROGRESS
AND PERSPECTIVES

Mortimer L. Mendelsohn
John P. Peeters
Mary Janet Normandy
Editors

JOSEPH HENRY PRESS
Washington, D.C. 1995

JOSEPH HENRY PRESS   2101 Constitution Avenue, N.W. Washington, D.C. 20418

This volume is primarily based on scientific papers presented at a workshop on the use and application of biomarkers held in Santa Fe, New Mexico, in April of 1994. The workshop was sponsored by the U.S. Department of Energy. Any opinions, findings, conclusions, or recommendations expressed in this volume are those of the authors and do not necessarily reflect the views of the U.S. Department of Energy or the National Academy of Sciences.

Library of Congress Cataloging-in-Publication Data

Biomarkers and occupational health : progress and perspectives /
  Mortimer L. Mendelsohn, John P. Peeters, Mary Janet Normandy,
  editors.
        p.   cm.
    Includes bibliographical references and index.
    ISBN 0-309-05187-8
    1. Biochemical markers. 2. Industrial toxicology. 3. Biological
monitoring. 4. Occupational diseases——Diagnosis. I. Mendelsohn,
Mortimer L. II. Peeters, John P., 1958- . III. Normandy, Mary
Janet, 1944- .
RA1229.4.B566   1995
615.9′02——dc20   94-37672
                                                              CIP

AFP6138

Printed in the United States of America

Preface

This book is primarily based on the presentations from the International Workshop on the Development and Applications of Biomarkers, Santa Fe, New Mexico, April 26-29, 1994. This meeting was sponsored by three Offices of the U.S. Department of Energy (DOE) — Environment, Safety and Health; Energy Research; and Environmental Management. The DOE was motivated partly by current general interest in the field of biomarkers and partly by recent legislation encouraging the Department to monitor exposed workers.

The word biomarkers can have many meanings, but in the present context it refers to measurable properties which indicate that a subject has a particular susceptibility, has had a particular exposure, or is demonstrating a particular response to some known or unknown external perturbation. Taken together, these properties of biomarkers offer an approach to monitoring individuals in an environmental or occupational setting, and thereby estimating future, current, or integrated risk. The development and application of such methods promises the possibility of protecting workers, documenting their individual exposure histories, identifying threatening work situations, ameliorating potential adverse effects, and ultimately assigning responsibility for any risk that may have accumulated.

As most of us have noted, this is not a new field; reliance on biomarkers extends back into antiquity, including observations of hot skin, cyanosis, or the visual manifestations of fatigue and fear. Similarly, many current medical approaches to physical diagnosis and laboratory measurement involve biomarkers of disease or function. However, with the recent rapidly expanding concern for environmental risk, coupled with elegant improvements in sensitivity and breadth of analysis, biomarkers today are taking on a dramatically new technological and analytical character. The ultimate goal of these developments is sensitivity down to background levels, insight into mechanisms of action for the full range of exposure, and specificity for an enormous range of environmental agents. The reader is referred to any of the existing excellent summaries of the uses of biomarkers in environmental and occupational health research including those by the National Research Council.

At issue is whether current biomarkers are sufficiently sensitive and reliable for application, whether further progress can be made in their development, and how, eventually, a functional suite of biomarkers could be deployed in a variety of contexts.

The record, as contained within, will show that considerable capability and applications already exist for various types of biomarkers. It also demonstrates the wide disparity of opinion about the feasibility of using such biomarkers constructively. Attitudes differ considerably depending on whether the observer is a scientist who develops biomarkers, an epidemiologist, a physician responsible for monitoring personnel, a worker, a workers' union representative, or a responsible government official.

The evidence indicates an enormous potential for future biomarkers. Some will be developed in response to concern about specific relevant toxins. Others will be based on known mechanisms of toxicity. The Human Genome Project should be the source of many more biomarkers as it progressively defines human variability, susceptibility, and response. It is just a matter of time and energy to develop these capabilities, as well as the parallel mechanisms both to apply them efficiently and to fit them properly into current medical and legal frameworks for appropriate application.

While our legal system and the occupational medicine community may not be quite ready to deal with all of the issues related to the broad-scale application of biomarkers for worker protection, the path toward realizing this goal is increasingly clear and well worth taking. This book serves as a record of where we now are and how we might proceed into the future.

Mortimer L. Mendelsohn, M.D., Ph.D.
Vice Chairman
RERF, Japan

Contents

STUDY DESIGN

CLEANUP WORKERS AND OTHER MEDICAL NEEDS

RECENT TECHNICAL ADVANCES IN BIOMARKERS RESEARCH

CASES IN POINT: MONITORING WORKER EXPOSURES
TO METALS

BIOMARKERS

AND OCCUPATIONAL HEALTH

Biomarkers and Occupational Health: Progress and Perspectives. 1995. Pp. 1-6

Introduction: The Role of Biomarkers in the Prevention of Occupational Disease

Paul A. Schulte

Biological markers of xenobiotic exposure, early disease or disease susceptibility are all potentially useful tools for occupational health research and practice (Schulte, 1993a). Too often, the application of these markers has been less than optimum because of the failure of investigators to be precise about the type of marker that is needed or the specific purpose for which it will be used. This paper presents a framework for considering and using biomarkers in the prevention of occupational disease.

Table 1 shows the classic hierarchy of measures in the prevention of occupational disease (Olishifiski and McElroy, 1971; Weeks et al., 1991). In this hierarchy, the major objective is primary prevention, which involves modifications of the work environment including substitution, engineering and administrative controls, changes in work practices, and the use of personal protective equipment. Biomarkers are generally not included within the framework of primary prevention. However, biomarkers could be used in studies or trials to test the effectiveness of primary prevention efforts.

It is only in the second part of the hierarchy, involving secondary prevention, where the focus is on monitoring the worker for evidence of exposure or for early disease, that biomarkers are specific components of the prevention framework. The implicit mandate behind this framework is to change the environment before you change the worker. Many of the ethical issues pertaining to the use of biomarkers in occupational research and practices will fall into place if that dictum is kept in mind.

Historically, as environmental disease was viewed through the eyes of the epidemiologist, it was possible to relate exposure to disease often without understanding much about the mechanisms of action or the intervening steps (Figure 1). This approach has been successful in that it has contributed significantly to the improvement of public and occupational health. However, this approach is less useful in situations with low levels of exposure, mixtures of exposure, intermittent exposure or, in some cases, for diseases with long latency periods. In these cases, there have been tendencies to misclassify people as to their exposures. Because clinical disease was the outcome in question, it was necessary to include individuals with clinical disease before etiologic research could be fruitful.

TABLE 1. HIERARCHY OF MEASURES IN THE PREVENTION OF
OCCUPATIONAL DISEASE

A.   Reduce Exposure (Primary Prevention)

        1.   Substitution

        2.   Engineering controls

        3.   Administrative controls

        4.   Work practices

        5.   Personal protective equipment

B.   Reduce Effects of Exposure (Secondary Prevention)

        1.   Medical monitoring

            A.   Pre-exposure screening

            B.   Detection of evidence of excessive exposure

                i.   Tests of body burden

                ii.   Tests of biological effect

        2.   Control of additive and synergistic exposures

C.   Reduce Effects of Disease (Tertiary Prevention)

        1.   Rehabilitation

        2.   Job re-entry

Further, one of the biggest weaknesses of the classic epidemiologic approach has been in accounting for host factors. Historically, epidemiologists accounted for host factors in terms of age, race, sex, and some behavioral factors. They have not really accounted for those acquired or genetic factors that can significantly influence the metabolism of toxicants.

Traditional Epidemiology

FIGURE 1. Traditional epidemiologic view of the relationship between exposure and disease.

There is now a set of tools to address these problems of exposure misclassification, late disease, and failure to account for host factors: biomarkers. Figure 2 presents the continuum of events between exposure and disease (NRC, 1987). Previously, epidemiologists measured ambient exposure; now it is possible to quantify with precision the amount of xenobiotic that gets into the person, i.e., the internal dose, and the amount that reacts with critical macromolecules. These amounts can be depicted by various biological indicators known as markers of exposure.

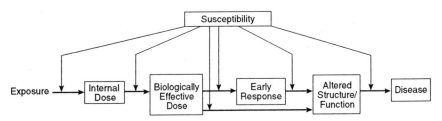

FIGURE 2. The relationship of biological markers to exposure and disease (Adapted from National Research Council, 1987).

At the other end of the continuum is disease. However, instead of waiting for clinical disease, it may be possible to trace back to the middle of the continuum in the natural progression of disease and start to identify preclinical or subclinical events that might be indicative of early disease. These events are referred to as markers of effect and ideally, they will make it possible to identify early warnings of severe conditions so that effective intervention measures can be initiated.

The markers in the center of the continuum are in a state of flux; they are indicators of biologic change. Depending on the state of the science and how they have been validated, these markers could indicate either exposure or biologic effect. These markers of biological change are useful tools, especially because many of them are non-specific indicators; they represent the integration of exposures from various sources or by various routes.

Superimposed on the multiple steps in the continuum is a new set of tools that are markers of susceptibility or predisposition. These are markers that indicate increased probability of proceeding from one step to the next. These are the markers that in some cases have the potential for being the focus of discrimination and other negative aspects of biomarkers that can emerge if they are misused.

In discussing biomarkers, it is important to distinguish the categories of exposure, effect, and susceptibility and their uses, strengths, and weaknesses. Markers of exposure, somewhat obviously, provide more detailed indications of exposure. A major limitation is that most of the markers currently in use only represent fairly recent exposure: days, weeks, months, in some cases, a few years. There are some biological dosimeters that may be able to depict historic exposures; but these have generally not yet been perfected. With markers of effect, it is possible to take previously homogeneous classifications of disease and distinguish subsets within them that might be related to different etiologies. With markers of susceptibility, there is an opportunity to distinguish high- and low-risk populations or even individuals.

Therefore, biomarkers can now be added to the hierarchy of prevention and should be thought of as supplementary to the traditional practices, not as a replacement (Schulte, 1993b). Biomarkers are not a panacea; they are an additional set of tools to be used with the traditional tools such as exposure assessments and the use of questionnaires to obtain information on confounding factors and risk factors.

Historically, occupational medicine has used biological markers. A factor that in part distinguishes this current generation of biomarkers from the past generation is the increased sensitivity that makes it possible detect xenobiotics at increasingly lower levels and at earlier stages in the natural history of disease. As with many technological advances, the ability to detect has far outstripped the ability to understand. Still these markers are potentially very powerful tools.

It is important for those who want to use these tools to consider the reasons for using a biomarker and what kinds of markers and study designs are appropriate for their purpose. Will they be used to measure exposure or effect? Will they be used to monitor a population? At that point, it is necessary to consider the specific biomarkers that are available and decide whether they are practical, reliable, and valid, and whether the user will know how to interpret the results.

Validity has a variety of definitions. To the laboratory scientist, the concept of validity involves knowing the characteristics of the marker and whether it can be detected under certain conditions. To the epidemiologist who wants to use these markers in field studies, that information is not enough. The epidemiologist also has to know whether this marker will vary in the population, whether it will vary in different subgroups, e.g., by age or by race, and what size sample would be needed.

Before an epidemiologist can use a given marker in an epidemiologic study or even before a clinician can use it for biologic monitoring, they have to know something about its variability in populations. To attain this kind of information essentially requires an epidemiologic study in its own right. Performance of these epidemiologic validation studies is where much of the emphasis on biomarkers is currently being placed.

To further clarify the discussion of biomarkers, Dr. Nathaniel Rothman of the National Cancer Institute and I have constructed a common sense matrix that might be useful. In general, we can consider a progression of study designs, starting with laboratory studies that develop the marker, and then a series of transitional studies that are intended to characterize the marker, taking it from the laboratory to the field. Then we consider two possible uses of the markers. One is in etiologic studies where instead of testing the marker, the marker is used as a dependent or independent variable in an epidemiologic study design. We can also consider applied uses of the marker in biological monitoring or medical screening programs. These are two uses that could apply to workers at nuclear waste sites or who are involved in cleaning up weapons plants.

No matter how biomarkers are used, there is a need to recognize the importance of notifying workers, not only of their own biomarker test results, but of the overall study results employing biomarkers. This is important because workers have the right to know how their own results fit into the general context of such tests. Moreover, it is important to understand and consider not only the scientific aspects of biomarkers, but also the social, political, cultural, and legal aspects. It is incumbent upon the scientists, regulators, and research agencies who are involved in the biomarker issue to participate in these discussions as well as in the discussions of the validity and the scientific merit of the individual markers.

REFERENCES

National Research Council. 1987. Biological markers in environmental health research. Environmental Health Perspectives 74:1-191.

Olishifiski, J. B., and F. McElroy, eds. 1971. Fundamentals of industrial hygiene. National Safety Council, Chicago, Pp. 439-480.

Schulte, P. A. 1993a. A conceptual and historical framework for molecular epidemiology. Pp. 1-44 in: P. A. Schulte and F. P. Perera, eds. Molecular Epidemiology: Principles and Practices, Academic Press, San Diego.

Schulte, P. A. 1993b. Use of biological markers in occupational health research and practice. Journal of Toxicology and Environmental Health 40:359-366.

Weeks, J. C., B. S. Levy and G. R. Wagner, eds. 1991. Preventing Occupational Disease and Injury. APHH Washington, Pp. 45-51.

NATIONAL AND INTERNATIONAL PERSPECTIVES

Biomarkers and Occupational Health: Progress and Perspectives. 1995. Pp. 9-19

The Current Applicability of Large Scale Biomarker Programs to Monitor Cleanup Workers

Mortimer L. Mendelsohn

The atom-bomb survivors of Hiroshima and Nagasaki are in many ways an ideal population for evaluating the application of biomarkers, or, as a more preferable term in this context, biological dosimeters. The survivors were exposed acutely to ionizing radiation in 1945 and have been studied intensively for health effects throughout the intervening 48 years. Chromosomal aberrations, the classical dosimeter of radiation effect, have been measured for decades, and in recent years, almost a half-dozen mutational dosimeters have been used as well. This population has probably been better studied from this point of view than any other in the world.

A good starting point is the data on glycophorin A mutations in erythrocytes. Glycophorin A is a sialoglycoprotein found exclusively and abundantly in the red-cell membrane. The gene for glycophorin A in humans has two equally prevalent alleles, M and N, whose gene products differ by two amino acids and are the basis for the M and N blood types. The glycophorin A somatic-mutation test is limited to the 50% of humans who are MN by blood type and, hence, normally express the M and N gene products on the surface of every red cell. The test uses flow cytometry and differently labelled fluorescent monoclonal antibodies to the M and N products to measure their content on the surface of each of several million red cells. Three mutant types are scored: the M0 and N0, which are those cells respectively lacking expression of either the N or M product but with normal amounts of the counterproduct; and the MM, which are those cells lacking the N product but with double amounts of the M product. M0 and N0 are interpreted as simple loss-of-function mutations in which the glycophorin A gene product from one chromosome 4 has been modified (lost or changed) to the point of being unrecognizable by the antibody, whereas the gene product from the other chromosome 4 is expressed normally. The M0 and N0 mutational events are statistically independent but otherwise identically likely events and can be averaged into a common estimate of the gene-loss endpoint. The MM appears to have two functioning and identical genes and probably represents reduction to homozygosity through some type of recombinational event. The background mutant frequencies are in the range of 5-20 per million cells for each of the three endpoints, depending on measurement method and age of the subject.

The atom-bomb survivors were one of the original populations to be investigated with the glycophorin A method, and from the beginning, they showed significant radiation dose-responses (Langlois et al., 1987; Kyoizumi et al., 1989). By now, well over 1,000 survivors have been studied, with results that are remarkably consistent (Akiyama et al., 1990, 1992). Figure 1 shows earlier data on the dose-response for N0 and M0 mutants in 300 survivors from Hiroshima. The near identity of these symmetrical but independent tests is clear.

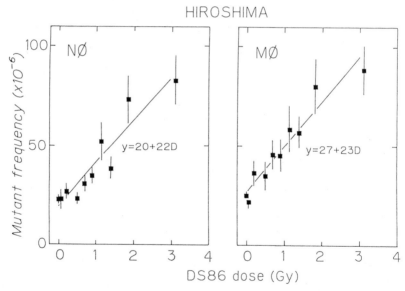

FIGURE 1. Glycophorin A gene-loss somatic-mutation response in atom-bomb survivors from Hiroshima. The data are means and 95% confidence limits for groups of roughly 30 survivors. Radiation dose, as estimated by the Dosimetry System 1986, is shown on the horizontal axes. Mutant frequency per million erythrocytes is shown on the vertical axes for each of two gene-loss endpoints, N0 and M0. The linear equation describes the mutant frequency per million cells, y, as a function of the zero dose intercept, the slope and D, the dose.

Figure 2 depicts the combined N0/M0 results plotted alongside the MM result, for the same Hiroshima population.

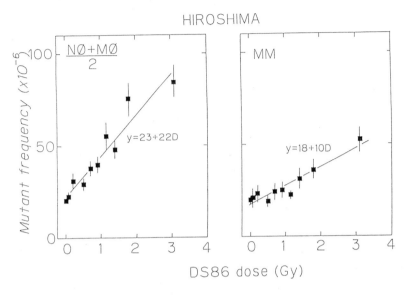

FIGURE 2. Glycophorin A gene-loss and recombinational response in atom-bomb survivors. Axes and notations are the same as in Figure 1. The left panel shows the average of the gene-loss data from Figure 1; the right panel shows the MM mutants. The slopes are significantly different, and both responses appear to be linear and without threshold.

The corresponding results for survivors from Nagasaki are shown in Figure 3. In effect, the gene-loss methods give mutational dose responses that are linear and without threshold, as would be predicted from the general experience with experimental cellular mutational systems. The recombinational endpoint behaves similarly but with about one-half the yield. The data by city differ in numbers of tests done, suggestively but not significantly in slope, and significantly in background level. Difference in slope could stem from systematic errors in the radiation yields of the two atom bombs, and particularly from their known differences in neutron output. The difference in the background or zero-dose mutant frequencies, which should be independent of the weapons, is totally unexpected and unexplained.

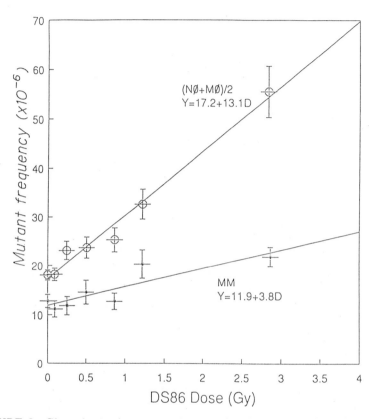

FIGURE 3. Glycophorin A response in atom-bomb survivors from Nagasaki. The axes, notations and analyses are the same as in Figure 2 except that here they are plotted in the same graph. The dose response from Nagasaki appears to be less steep than the Hiroshima response. However, the only significant difference between the cities is the background level for N0 + M0/2 mutants.

Apart from the background difference, these are surprisingly tidy results for a radiation effect that is now 48 years old. Such data encourage the thought that biological dosimeters could be helpful in situations where the physical or chemical doses are unavailable or uncertain. Before we become too optimistic about this concept, however, there are some important limitations. First there is the problem of *individual variability* with the glycophorin A assay. The assay is reasonably stable for repeat measurements on an individual; its standard deviation is only slightly larger than expected for counting errors. However, as shown in Figure 4, the scatter of measurements among atom-bomb survivors is large; the scatter of individual responses for any dose almost covers the entire range of the test.

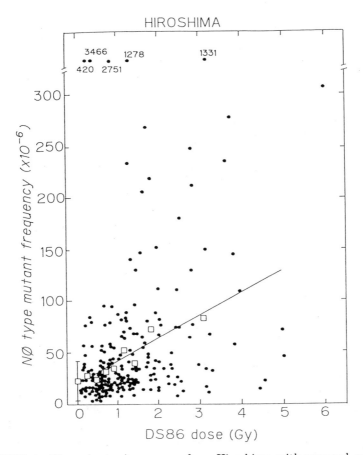

FIGURE 4. Glycophorin A response from Hiroshima with grouped and ungrouped data superimposed. These are the same format and data as in the left-hand panel of Figure 1. Note the dramatic contrast between the individual data and the grouped data, reflecting the large variation between individuals.

Part of this scatter must be due to the major uncertainty in radiation dose assigned to each survivor. However, we believe that most of the noise is fundamental to the biology of the test and is due to the limited number of stem cells that are available to record the initial mutational event. Note that a target population of approximately 10^5 cells is needed to score an average of one mutation, and well over 10^6 cells to estimate a reasonable frequency of mutation. A human, particularly one who has just been irradiated, has barely enough hematopoietic stem cells to measure a continuous response with this test. We estimate that a combination of about 10 people is needed to provide enough stem cells to make the test statistically reasonable. Such averaging is fine for the abundant and easily dose ranked atom-bomb survivors but would present great problems either for the relatively few people living around a toxic well, for example, or for populations for which no prior dosimetry allows sensible grouping of the subjects. In short, the glycophorin A test, useful as it is for large, ordered populations, cannot function as an individual dosimeter. To the extent that we understand the problem, we know that this shortcoming cannot be corrected by multiple measurements on the same individual or by counting more cells. Apart from the grouping of like people into a combined measurement, the only other potentially helpful strategy is to combine the test with other mutational endpoints into a composite estimate per person. However, this possibility has yet to be explored.

Another limitation is seen in the corresponding results for *lymphocyte-based mutation tests* on the atom-bomb survivors (Hakoda et al., 1988; Kushiro et al., 1992; Kyoizumi et al., 1992). Three contemporary lymphocyte assays have been tested: an HPRT assay, an HLA assay, and a TCR assay, each representing a specific gene suitable for mutation testing in human peripheral-blood lymphocytes. The HLA assay shows no evidence of dose response, whereas the HPRT and TCR assays give marginally significant responses that are a minute fraction of the slope one would expect from experiments on cells irradiated *in vitro*. The TCR and HPRT assays respond briskly in people with recent radiation exposures. Based on this and other evidence, we suspect that the mutational signal in these lymphocyte systems decays with a half-life of several years, presumably due to selection against the mutant phenotype and/or to cellular turnover in which the initially mutated cells are replaced from a pool with relatively low mutant frequencies.

Let us now review the issues that will arise in applying the biomarkers of exposure and effect.

Temporal Effects in Biodosimetry

Mutational tests reach backward in time differently. The glycophorin A test seems to reflect exposure for most of a human lifetime. This is also true for measurements with chromosomal reciprocal translocations in lymphocytes. However, the tests using lymphocyte mutation systems and chromosomal dicentrics operate well only over a span of several years. DNA adducts, a collection of biomarkers reflecting chemical exposures, have short time spans, ranging from fractions of a day to perhaps a week. Temporal effects also involve differences in how long it takes an assay to become positive after an acute exposure. Adduction and chromosome aberration are expressed rapidly, whereas lymphocyte mutation may take a few weeks for full expression. The glycophorin A system depends on the mutant red cells entering the peripheral-blood pool, and hence, roughly two months elapse before the assay reaches 50% levels. This delay is a severe limitation for acute exposure situations.

The Specificity of Biomarkers

Do biomarkers work for all mutagens, carcinogens and toxins? The chromosomal and gene-mutational markers should respond to essentially all mutagens with access to the relevant target cells. These assays also are likely to respond to half of the known carcinogens, judging from the extensive literature on mutational testing as a predictor of carcinogenicity (Mendelsohn, 1988). Promotional and otherwise nonmutational carcinogens, representing the other half, will not be detected, including the dioxins and related chlorinated compounds that often appear on inventories such as those found at cleanup sites. Likewise, nonspecific toxins should not affect these systems. Many of the detection methods for DNA adducts are adduct-specific; to use these systems effectively one must have prior knowledge of the chemical of concern. Also, adduction is a step toward mutation but is not mutation *per se*, because many adducts are repaired without ever causing mutation. Thus, adduction is an indicator of mutagen exposure, whereas the chromosomal and gene-mutational assays are indicators of both mutational exposure and effect.

Another aspect of specificity is whether a positive response with a bioassay necessarily indicates an occupational exposure. As is always the case in occupational epidemiology, tobacco use and other lifestyle factors, prior occupational exposures, and innate hypersensitivity must be evaluated as potential confounders of a positive response. Mutational assays respond modestly to tobacco and dramatically to the rare inborn errors of DNA repair. The effects on the assays of other lifestyle, genetic, and occupational factors are not yet known, but one can predict complications from diet, medical radiation, and medications. The unknown source of difference in the backgrounds of Hiroshima and Nagasaki (described above) may well be caused by such confounders. Another example is the unpublished finding that the TCR mutant frequency in the Japanese studies is higher in females than in males. We will probably find many other such surprises within the large background variation that characterizes the mutational biomarkers.

The Sensitivity of Biomarkers

The issue of sensitivity is extremely important and may make or break the application of these methods. The dose responses for glycophorin A mutations range over several gray, while occupational exposures are more likely to be in the range of 50 mGy or less. I know of no detailed statistical estimates of the point along these dose responses where the data first become significantly different from background. Judging from the confidence limits of the clusters of data, however, the sensitivity limit is roughly 500 mGy. Thus, the measurement of hundreds of atom-bomb survivors (with all the concomitants of uncertainty in radiation dose and 40-year latency) has a sensitivity about 10-fold above the likely upper range of occupational exposure. This is for ionizing radiation, the agent for which we have the best understanding of how these tests behave. For chemical mutagens, the situation is less clear, although the data for chemotherapeutic agents suggest results similar to those for radiation. For environmental pollution, the sparse data are limited to a few severe exposure conditions and marginally positive results. For example, the data on the Japanese atomic bomb survivors involve much larger environmental exposures than one would expect to encounter among the workers involved in the cleanup of nuclear weapons sites.

Validation of Biomarkers

Validation of biomarkers takes two forms: establishing reasonable dose-response relationships for representative agents expected in the field; and determining the behavior of assays directly by applying them to occupational groups. Generation of atom-bomb-type dose responses for the dozens of chemical agents found in the United States nuclear sites inventories of environmental contaminants would be an enormous challenge. Dose responses require human populations with known exposures to single agents, and such populations are extremely unlikely to exist. Learning by doing is the obvious alternative, but this too presents a great challenge. Consider the situation in which a group of cleanup workers is found suddenly to have positive responses in a screening system using a battery of biomarkers. The almost certain response will be the immediate cessation of the operation, followed by a detailed evaluation of the procedural deficiencies that might have allowed such an exposure. No administrator would dare to continue or to duplicate the exposure, making it impossible either to corroborate the initial result or to refine the exposure conditions. Administrative and legal considerations necessarily prohibit using this situation as a continued learning experience.

All of the above issues of temporal effects, specificity, etc., raise serious questions regarding the utility of biomarkers in a large-scale setting, such as the populations involved in the cleanup of the United States nuclear weapons complex. Situations may occur where occupational exposures are truncated beneficially because of biomarker response. But, given the current uncertainties in dose-response and specificity of response of the known biomarkers of exposure and effect, we will probably never know which situations are statistical or technical artifacts and which are bona fide toxic exposures. Nor is the wholesale application of these methods in the field likely to contribute to our knowledge of the methods *per se*. Perhaps the one hope for this technology in this setting is that parallel research into biomarkers will be stimulated by the United States needs and will eventually provide the desired information.

RECOMMENDATIONS

I would like to make the following recommendations about biomarkers in relation to new United States mandates to monitor cleanup workers:

- Postpone, if possible, the large-scale application of biomarker testing of occupational groups until the prospects for benefit are better defined.

- Meanwhile, promote the application of biomarkers by:

- Development of additional mutational and nonmutational biomarkers of exposure and effect.

- Identification of suitable populations for testing of biomarkers, such as:

    - Atom-bomb survivors

    - Chemotherapy patients

    - Radiotherapy patients

    - Victims of chemical and radiation accidents

    - Populations with high environmental exposure

- Definition of general protocols for identification and validation of biomarkers.

- Coordination of multiple biomarkers into appropriate batteries.

- Analysis and improvement of biomarker sensitivity.

- Determination of the functionality and cost-benefit aspects of biomarkers in the context of the United States needs.

ACKNOWLEDGEMENTS

This publication is based on CR 1-94 concerning research performed at the Radiation Effects Research Foundation (RERF). RERF is a private, nonprofit foundation funded equally by the Japanese Ministry of Health and Welfare and the United States Department of Energy through the National Academy of Sciences.

REFERENCES

Akiyama, M., Y. Kusunoki, S. Umeki, Y. Hirai, N. Nakamura, and S. Kyoizumi. 1992. Evaluation of four somatic mutation assays as biological dosimeters in humans. Pp. 177-182 in Radiation Research: A twentieth-century perspective. W. C. Dewey, M. Edington, R. J. M. Fry, E. J. Hall, G. F. Whitmore, eds. Vol II, Academic Press, New York.

Akiyama, M., S. Kyoizumi, M. Hakoda, N. Nakamura, and A. A. Awa. 1990.

Studies on chromosome aberrations and HPRT mutations in lymphocytes and GPA mutation in erythrocytes of atomic bomb survivors. Pp. 69-80 in: Mutation and the Environment. Part C: Somatic and heritable mutation, adduction and epidemiology, Progress in clinical and biological research. Vol 340C M. L. Mendelsohn and R. J. Albertini, eds. Wiley-Liss, New York.

Hakoda, M., M. Akiyama, S. Kyoizumi, A. A. Awa, M. Yamakido, and M. Otake. 1988. Increased somatic cell mutant frequency in atomic bomb survivors. Mutation Research 201:39-48.

Kushiro, J., Y. Hirai, S. Kyoizumi, Y. Kodama, A. Wakisaka, A. Jeffreys, J. B. Cologne, N. Nakamura, and M. Akiyama. 1992. Development of a flow-cytometric HLA-A locus mutation assay for human peripheral blood lymphocytes. Mutation Research 272:17-29.

Kyoizumi, S., N. Nakamura, M. Hakoda, A. A. Awa, M. A. Bean, R. H. Jensen, and M. Akiyama. 1989. Cancer Research 49:581-588.

Kyoizumi, S., S. Umeki, M. Akiyama, Y. Hirai, Y. Kusunoki, N. Nakamura, K. Endoh, J. Konishi, M. Sasaki, T. Mori, S. Fujita, and J. B. Cologne. 1992. Frequency of mutant T lymphocytes defective in the expression of the T-cell antigen receptor gene among radiation-exposed people. Mutation Research 265:173-180.

Langlois, R. G., W. L. Bigbee, A. Kyoizumi, N. Nakamura, M. A. Bean, M. Akiyama, and R. H. Jensen. 1987. Evidence for increased somatic cell mutations at the glycophorin A locus in atomic bomb survivors. Science 236:445-448.

Mendelsohn, M. L. 1988. Can chemical carcinogenicity be predicted by short-term tests? In: Living in a Chemical World. Vol 534, Annals of the New York Academy of Sciences.

Biomarkers and Occupational Health: Progress and Perspectives. 1995. Pp. 20-24

Biomarkers - A Perspective from the Commission of the European Communities

Jaak Sinnaeve and Ken H. Chadwick

The Radiation Protection Research Division of the Commission of the European Communities (CEC) conducts work in the fields of radiation sensitivity, radiation carcinogenesis, and modelling of radiation carcinogenesis. In these fields, it is critical to develop and identify indicators of exposure, sensitivity, and susceptibility to radiation. This paper presents a review of the European radiation protection program and of the need for a concerted international effort in biomarker development. Our discussion is limited to radiation because our research program is not concerned with chemical exposures.

Indicators of Exposure to Radiation

In the past, our program supported work on biological dosimetry using dicentric chromosomal aberration induction in blood lymphocytes. This method has been successful for doses in excess of 20 cGy when measurements are made within a few days of exposure. Work using micronuclei induction has also been supported, but this method is less useful because of variations in the background levels in different individuals and because of a saturation of the dose-effect relationship at higher doses. The technique of chromosome painting allows the use of stable translocation-type aberrations that remain in the lymphocytes and might provide a cumulative measure of exposure. We have begun to examine how useful these aberrations might be for measuring accumulated dose. One problem will be to determine the extent to which other factors, such as chemical exposure, viral infection, or aging, can influence the occurrence of the translocations.

Individual Radiation Sensitivity

The CEC supports a well coordinated research effort to study DNA repair and the molecular basis of radiation sensitivity. Some work has also been done on the use of different techniques to measure individual radiation sensitivity. In 1988, together with the Atomic Energy Control Board of Canada Ltd., the Commission organized a workshop on methods for determining differential radiosensitivity in humans. At that time, there was no clear method available that could unambiguously identify radiation sensitivity in individuals, unless it was extreme sensitivity such as that associated with Ataxia Telangiectasia (AT). A group of researchers from the United States reported that they could identify AT heterozygotes who are suspected to be slightly radiation-sensitive using the measurement of G2 chromatid break repair. They also could identify cancer-prone individuals with the same test. Only one other group has, as yet, been able to reproduce these results. Perhaps in the near future, there may be a more definitive answer on the usefulness of this G2 assay.

More recently, an assay, referred to as the Comet assay, has been developed to look at DNA damage in individual cells, and it has been suggested that this may also be a useful test of radiation sensitivity. However, there are considerable difficulties in quantification of this assay.

Susceptibility to Radiation Induced Cancer

It is known that some people have a predisposition to develop certain forms of cancer. This conclusion is based, in part, on the work of Knudson (1971) on inherited childhood cancers and also on the ability of molecular biologists to study tumor suppressor genes. There is reason to believe that these predisposed people have greater susceptibility to the induction of cancer by radiation and other mutagenic agents. Indeed, in hereditary retinoblastoma patients at increased risk for osteosarcoma, it appears that risk is especially high within radiation-treated patients (Eng et al., 1993).

In the near future, it may be possible to use molecular biological techniques to identify individuals who are predisposed to certain cancers and will probably be susceptible to radiation-induced cancer. And as a corollary, it may then also be possible to identify individuals who have greater resistance to radiation-induced cancers.

Radiation "Signature"

The idea that radiation damage may leave a unique "signature" in cells and that this signature can be used to determine whether the tumor was actually caused by radiation is appealing. However, while such signatures might be particularly clear for chemical mutagens which only cause very specific base changes in the DNA, it is unlikely that the same will be the case for radiation. The initial damage induced by radiation includes a wide variety of DNA alterations, and the damage left in the tumor cell would be the consequence of the repair processing of that damage. Thus, in terms of a signature, it may be more realistic to look for typical mutations induced by radiation. Radiation-induced mutations often involve large deletions; however, recent work reported at a workshop in The Netherlands suggests that, as methods to detect small changes in DNA become available, it will be seen that mutations induced by radiation also include the smaller deletions and even "point" mutations (Okinaka et al., 1994).

Ethical Considerations

Because it may soon be possible to identify individuals predisposed to certain kinds of cancer and possibly at increased risk for radiation-induced cancer, we are faced with the problem of how to deal with this situation fairly in relation to potential exposure.

Radiation protection philosophy is generally oriented toward risk reduction and limitation and already takes into account the presence of more sensitive groups in the population. However, these considerations are based essentially on exposure pathways and on age rather than on cancer susceptibility and radiation sensitivity. This philosophy will need to be revisited in the near future, at least from the point of view of cancer susceptibility. There have been some interesting comments relevant to that debate and all of the ethical problems that it involves (Kagan, 1993; Maddox, 1993; Muller-Hill, 1993).

However, many confounding factors could make the debate even more complex. For example, the Moolgavkar and Knudson (1981) model for carcinogenesis proposes a pathway to malignancy that involves two mutational events separated in time so that the cell(s) carrying the first mutation can multiply and thus increase the number of target cells for the second mutation to induce the malignant cell(s). Spontaneous cancers presumably arise because some environmental agents can cause, within a lifetime, the two mutational events leading to malignancy. Radiation, in effect, increases the spontaneous rate of both mutational steps and should really be seen as "interacting" with the other environmental mutagenic agents. In this sense, radiation is potentially one of the contributory agents to the cancer, and it is impossible to separate the radiation effect from the effects of the other agents. In the case of smoking, it is clear that lifestyle may influence the "spontaneous" level of mutations and thus the level of the radiation interaction.

It seems likely that people inheriting a predisposition to cancer will, in many cases, be susceptible to radiation-induced cancer. It seems reasonable to advise these people not to accept employment that involves exposure to radiation, or any other mutagenic agent, and also to advise them on their lifestyle.

Combining our current knowledge of genetic predisposition in humans and experimental studies showing that some mouse strains are resistant to radiation-induced cancer might suggest that some proportion of the human population may be "immune" to radiation and will probably never get radiation-induced cancer. However, this is probably an erroneous assumption. As long as radiation can induce the two mutational events in the Knudson-Moolgavkar pathway that lead to malignancy, there is no reason why radiation might not eventually induce cancer in every exposed person.

CONCLUSIONS

Current advances in the wide range of topics that involve biomarkers and their use in radiation protection suggest that we may soon be able to avoid unnecessary exposures in susceptible people and to assess more critically the risk of exposure in others. These advances may lead to some revisions of radiation protection policies, but we do not believe that this will lead to the need for significant revisions of the ethical codes. It may never be possible to use molecular techniques to conclude with certainty that a tumor was induced by radiation.

REFERENCES

Eng C., F. P. Li, D. H. Abramson, R. M. Ellsworth, F. L. Wong, M. B. Goldman, J. Seddon, N. Tarbell, and J. D. Boice. 1993. Mortality from second tumors among long-term survivors of retinoblastoma. Journal of the National Cancer Institute 85:1121-1128.

Kagan, R. M. 1993. Justice is not egalitarian. Nature 363:578.

Knudson, A. G. 1971. Mutation and cancer: statistical study of retino-blastoma. Proceedings of the National Academy of Sciences (USA) 68:620-623.

Maddox, J. 1993. New genetics means no new ethics. Nature 364:97.

Moolgavkar, S. H., and A. G. Knudson. 1981. Mutation and cancer: a model for human carcinogenesis. Journal of the National Cancer Institute 66:1037-1052.

Muller-Hill, B. 1993. The shadow of genetic injustice. Nature 362:491-492.

Okinaka, R. T., S. L. Anzick, and W. G. Thilly. 1994. Denaturing gradient gel electrophoretic analysis of specific exons of the HPRT gene from X-ray induced mutant populations. Pp. 151-156 in Molecular Mechanisms in Radiation Mutagenesis and Carcinogenesis. K. H. Chadwick, R. Cox, H. P. Leenhouts and J. Thacker, eds. EUR 15924 European Commission, Luxembourg.

Biomarkers and Occupational Health: Progress and Perspectives. 1995. Pp. 25-34

The Role of the NIEHS in the Development of a National Program for Environmental Health Science Research

Ken Olden

As Director of the National Institute for Environmental Health Sciences (NIEHS) and the National Toxicology Program (NTP), I have visited the various Federal agencies that are involved in environmental research. The purpose of these visits was to ensure that as we developed our priorities, there would be no duplication of the activities going on in the Centers for Disease Control and Prevention (CDC), the Environmental Protection Agency (EPA), the Occupational Safety and Health Administration (OSHA), the National Institute for Occupational Safety and Health (NIOSH), or the National Center for Toxicological Research. We visited the Chemical Industry Institute of Toxicology and several major chemical firms, as well as the Department of Energy (DOE), the Food and Drug Administration (FDA), and the Department of Defense (DOD). We contacted representatives of labor unions, environmental groups, the public, and academicians. It was important to get an idea of the perceived problems in environmental health science research; to hear the criticisms of the way NIEHS had operated in the past; and to hear opinions on what should be done in the future.

A most difficult task has been to convince both policymakers and the American people that human diseases and dysfunctions are related to both genetic and environmental factors. Although that connection would appear to be self-evident, the budget of risk assessment research in the federal government indicates that public sentiment for the environment has not translated into dollars for environmental health science research.

Therefore, three challenges faced me as director of NIEHS. The first challenge was to increase the visibility of the institute; the second was to increase the budget. And since I thought both of these goals could be accomplished by increasing the quality of the science, the third challenge was to increase interactions with agencies such as DOE, EPA, and CDC.

A number of collaborative interactions are currently under way. For example, in collaboration with DOE, EPA, and other agencies, NIEHS recently sponsored a conference on environmental equity in Crystal City, Virginia. During the same week, President Clinton released his Executive Order on environmental justice.

NIEHS is also collaborating with DOE on electromagnetic field (EMF) research and on the training of workers to handle hazardous materials. These ongoing collaborations with various other federal agencies, as well as with academia and industry, represent the new paradigm for government research.

NIEHS will become involved with NIOSH to develop collaborative interactions in order to make certain that we do not duplicate efforts and that all the important public health issues related to the environment are being addressed by at least one of our agencies.

It is obvious that human risk assessment research could be greatly improved. If we are going to make better decisions in terms of the impact of the environment on human health, we need better science.

Challenges in Assessing Environmental Risk

One of the main challenges in environmental health sciences research is the accurate assessment of human risk. It is tremendously costly to regulate human exposures to some of the products of modern technology. The key is to determine the level of regulation required based on the true hazard of environmental exposures to human health. Over-regulation is wasteful in terms of dollars spent and the loss of benefits from products. Under-regulation is costly in terms of human health, welfare, and productivity. We must seize every opportunity to understand the basis of the human diseases that result from exposures to environmental agents.

Accurate assessment of human risk resulting from environmental exposures requires three major kinds of information:

- **Knowledge of exposure levels and rates of exposure.** This is a major limitation in human population studies. Reconstructing dose histories in populations is extremely difficult due to a lack of historical information. Even if accurate external measurements exist, individual doses are usually impossible to reconstruct. Biomarkers of exposure are now being developed and validated for specific chemical exposures.

- **Hazard identification.** This is usually accomplished through epidemiologic studies, which are relatively weak in their abilities to demonstrate causal relationships between exposures and health effects, or through bioassays in laboratory animals using individual dosing regimens and biological models of health effects followed by extrapolation back to human exposures. The extrapolations from high to low dose and between species (using size scaling and accounting for metabolic differences) also leave much to subjectivity and uncertainty. Biomarkers indicating early disease stages are needed, and NIEHS has solicited grant applications in this area of study.

- **Knowledge of mechanisms of action.** Understanding *how* a substance exerts its toxic effect is the key to removal of the large uncertainty factors currently being used in today's risk assessment procedures. Confidence in the applicability of experimental findings in laboratory animals to the true human experience can increase tremendously if the mechanisms of disease induction are known to be the same. Likewise, information that mechanisms in test animals are not relevant to humans reduces the concern that would result from a positive bioassay finding. This is why mechanistic studies must be included in the future of the National Toxicology Program and why the NIEHS has been reorganized to integrate basic laboratory investigations with the studies of the NTP. Biomarkers can be guideposts in such studies.

The Role of NIEHS

NIEHS needs better science communication with the American public. The format of our new journal, *Environmental Health Perspectives*, has been changed to resemble that of the *New England Journal of Medicine*. Our goal is to communicate in language that lay-people, lawyers, physicians, and school teachers can understand.

In addition, NIEHS is putting great emphasis on prevention and intervention. Our theme is good science for good decisions through communication, intervention, and prevention.

Currently, the opportunities in environmental health science research have expanded dramatically. This is mainly due to the revolution in cell and molecular biology. With developments in analytical chemistry, computational chemistry, and molecular biology, the tools are now available to answer a lot of fundamental questions.

A major challenge in environmental health science research is to establish causal associations. We have not been able to demonstrate strong causal associations using clinical epidemiological studies, but now using molecular techniques, we can demonstrate and measure exposures, biological effects, and susceptibility. Therefore, we can now show much stronger likelihood of causal associations.

Reliable biomarkers of both exposure and of pre-clinical disease can eliminate uncertainty during periods of latency between exposure and disease. Those long latency periods have previously made it difficult to establish causal associations in human populations. The field of molecular epidemiology attempts to identify *biomarkers of exposure* such as metabolite complexes in easily sampled tissues such as blood, *biomarkers of effect* such as changes in enzyme systems or gene structure and products, and *biomarkers of susceptibility* such as the existence of genes or other measurable biologic indices of susceptibility to the specific environmental factor.

Transgenic animals are an example of the new tools at our disposal. Susceptibility genes can be taken out of human cells and put into animals such as mice. These animals can be subjected to specific environmental agents in order to determine whether breast cancer susceptibility genes, bladder cancer susceptibility genes, colon cancer susceptibility genes, or lung cancer susceptibility genes render these animals more susceptible to certain kinds of carcinogens. Homeobox genes, which are believed to be involved in development, can be inserted into animals to determine if certain kinds of exposures render these animals more or less susceptible to certain kinds of environmental agents.

New Research Priorities

There are several things that should now be addressed. First of all, methodological research should be conducted to develop new approaches and new models to improve extrapolation from laboratory animals to humans and to improve extrapolation from high dose to low dose. A number of models can be used, for example, transgenic animals or transgenic cells.

In addition, we need models that represent other disease end-points. NIEHS is now soliciting ideas from the extramural community to help develop new models that will answer some of the fundamental questions so that we can begin to do more biologically-based risk assessment. Human cells can be used in order to obviate the necessity for extrapolation from rodents to humans. We can begin to use animal studies, in some cases, because of the availability of biomarkers.

Additionally, we need new and powerful models for non-cancer end points. Environmental health science research and risk assessment research originated in the federal government about 25 years ago in response to public concern about cancer. In the past, NIEHS has focused mainly on cancer as a disease end-point. However, it is clear that the environment plays a major role in the etiology of a number of other diseases and dysfunctions. Asthma and other pulmonary dysfunctions, neuroendocrine disorders, Parkinson's disease, Alzheimer's disease, osteoporosis, and endometriosis are all dysfunctions that are probably affected by environmental exposures, and we need to begin to address those relationships in a serious way.

NIEHS, unlike the other NIH institutes, deals with a broad spectrum of human diseases and dysfunctions, and it is the only institute with that broad responsibility. We also need to improve our capacity to predict the toxicity of chemical agents for which there is very little toxicity information. Estimates vary, but there are currently about 62,000 chemicals in commercial use in the United States, and about 1,500 new ones are introduced every year. It would be impossible to test every chemical, but there should be an increase the number of chemicals tested. One way of improving our capacity to predict toxicity is the use of "functional toxicology." Functional toxicology could involve, for example, the study of an array or several arrays of transgenic cells. In a matter of weeks, hundreds of chemicals could be screened for function. The National Toxicology Program has tested more than 450 chemicals, and we have learned a lot about structure-activity relationships. We now have a database that can be used to begin to generate computer models to make the kinds of predictions that are necessary.

In 1993, NIEHS conducted a small study to see how well we could predict the carcinogenicity of chemical substances. The results were best for chemicals at each end of the spectrum, those that are definitely carcinogenic and those that are definitely not carcinogenic. The substances in the middle could not be predicted with complete reliability; overall accuracy was only about 80 percent. However, this level of predictability resulted in part because there were certain gaps in our information. Now we know what data are needed on which types of chemical substances. Functional toxicology procedures can be used to indicate which chemicals should be entered into the long-term, two-year bioassay.

We estimate that approximately 100 chemicals can be screened in a matter of weeks. The two-year rodent bioassays cost from $2 million to $4 million per year. Therefore, it is important to make certain that the chemicals that are tested are those that cannot be predicted by some other reliable means.

Another important goal is to incorporate cutting-edge technology, such as cell and molecular biology, into risk assessment research. Historically, even within NIEHS, epidemiology, toxicology, and molecular biology were separate entities. There was not the kind of interaction that would result in synergy between the three groups. Epidemiologists were not taking advantage of the cutting-edge technologies in cell and molecular biology that existed even within our own Institute. The situation was similar for toxicologists. This is not an efficient way to operate. It is important for the basic scientists at NIEHS to address the research goals of the National Toxicology Program. For example, if suppressor genes are important in the mechanism of toxicity or carcinogenesis of a given substance, it would be logical for molecular biologists to participate in the design of targeted, specific, applied kinds of research.

That is exactly the approach that NIEHS has taken. The molecular biologists who were conducting basic research have begun to focus on the goals of the basic research program of the Institute mission. NIEHS was created to address the impact of environmental exposure on human health. Therefore, our basic research program is not similar to that of the National Institute of General Medical Sciences of the National Institutes of Health. Our research has to be specifically targeted toward diseases that are caused or exacerbated by factors in the environment.

Within NIEHS, those three groups, toxicologists, molecular biologists, and epidemiologists, are now working very closely together. In epidemiology, we are developing the molecular tools that are needed to do screening for phenotyping in communities in order to conduct exposure assessments. Therefore, it is imperative that state-of-the-art cell and molecular biology becomes incorporated into our toxicology and epidemiology activities.

The other area that should be strengthened at NIEHS is mechanistic studies. Historically, NIEHS has been involved in descriptive toxicology and descriptive epidemiology. Mechanistic studies can be used to better extrapolate from high dose to low dose, from rodents to humans.

A risk assessment group has been created at NIEHS that considers what kinds of information would be required to do intelligent human risk assessment on a chemical or physical agent that is toxic or carcinogenic. For example, if a pathway that metabolizes a given chemical to its carcinogenic state is demonstrated to exist in both humans and rodents, then it is likely, although not certain, that the agent will be carcinogenic in humans. If the metabolic pathway does not exist in humans, then it is likely, but not certain, that the agent is not going to be carcinogenic or toxic to humans.

As more is understood about mechanisms, dose-response relationships, pharmacokinetics, and excretion rates, we can better predict the likely exposure level and outcome in humans. Our animal studies are usually conducted at a maximum tolerated dose in addition to two other doses. All of the tested doses are high compared with actual human exposures except in special circumstances such as in occupational exposures, which might in some cases approach the doses used in NTP test assays. The important question is whether these lower levels of exposure cause human disease or dysfunction. The way to answer that question is through mechanistic studies, pharmacokinetic studies, and the development of dose-response relationships.

However, there is still a necessity to determine the potential effects at a high dose because occupational populations and sub-populations, such as members of communities that are located in hazardous environments, may be exposed to very high doses.

Of course, we need to develop biomarkers: biomarkers of exposures, effect, and susceptibility. It is very clear that susceptibility is going to be a very important and different issue. Although we understand the molecular basis of how some chemicals act, we generally do not yet have a clear understanding of what accounts for differences in susceptibility. DNA repair enzymes, differences in patterns of detoxifying enzymes or differences in rates of transport or excretion are just some of the factors involved.

Studies with animals demonstrate that susceptibility can vary as much as 1,000-fold. In regulation, a factor of 10 or, at most, 100 is used to account for sensitivity differences among individuals in a population. The observations in animal studies suggest that, under current guidelines, some people are likely to be exposed to doses of a given agent that are harmful to their health. Therefore, we need to develop the ability to phenotype individuals and determine or estimate the risks for individuals and/or populations and sub-populations. Differences in susceptibility is clearly a very important area of research.

There should also be greater emphasis on cross-species studies. We must begin to look at metabolic pathways, how chemicals are processed among various species, starting with rodents or even lower animals and progressing to humans. That will require the introduction of clinical studies.

Another important and difficult issue is exposure to mixtures. It is difficult enough to establish a causal association between one agent and a disease or a dysfunction; understanding the interaction between exposure to mixtures and disease is much more complicated. However, in real life, humans are exposed to mixtures, not to individual agents one at a time. Currently, NIEHS cannot put a lot of resources into the study of mixtures, but we must begin to try to understand and develop the means and the tools to explore mixtures. Exposure to mixtures is an extremely important issue to socioeconomically disadvantaged populations because of where they live and work. Therefore, we cannot continue to say that mixtures are such a difficult issue to deal with that we cannot do research. Some research on mixtures is already ongoing in a limited way in relation to the carcinogens in tobacco smoke. Mixtures is an area of research that must certainly be strengthened.

Another issue is the possibility of increased susceptibility of certain subgroups within the population. Individuals who are malnourished are probably totally different in terms of their responses to toxic substances. In addition, children are treated in regulation as if they are miniature adults. Children are not miniature adults; they are biologically very different from mature adults and we need to recognize that. Also, the elderly might be compromised immunologically and may thus be more susceptible to toxic agents. We need to understand these relationships and addresses these issues when regulations are developed.

The Importance of Prevention

The individual institutes of the National Institutes of Health focus on treatment and rehabilitation. Many people cannot afford these treatments because they are based on high technology and they simply cost too much. Every effort needs to be made to bring down health care costs. In order to achieve that, much greater emphasis should be put on prevention. At NIEHS, everything that we do is related to prevention. Environmental health science is about prevention. It begins with identifying the hazards and conveying that information to regulatory agencies;

then regulatory policies are established. Another approach, which has not been addressed very often, is to develop intervention strategies that do not depend on expensive and highly technical approaches.

The current impetus in the United States is for reducing health care costs so that everyone can afford health care coverage. The nation cannot afford to rely on high technology-based expensive treatments. Environmental health science research can develop cost-effective, straightforward, non-high technology intervention strategies. For example, if a xenobiotic such as dioxin interacts with a receptor, then it is possible to develop an antimetabolite that will prevent dioxin from binding to that receptor. Information about the way an agent binds, for example chemicals with estrogenic activities such as PCBs, makes it possible to design antimetabolites to prevent disruption. In fact, we simply need to reverse what drug companies do in order to develop a therapeutic treatment. The technologies and methodologies are well established. We simply need to apply them.

We can also develop low cost intervention strategies that will work even after an individual has been exposed to an environmental agent. It is probably very easy to use interventions in these cases because there is often a long lag period between exposure and clinical manifestation of a disease. That long lag can be anywhere from 20 to 40 years. Agents to which we were exposed during our childhood can impact our lives at ages 65, 70 or 80. With appropriate biomarkers, we could phenotype and track people during that long lag period. We would know who has been exposed, who has been affected, and who is likely to develop cancer, Parkinson's disease, Alzheimer's disease, etc. Then we could begin appropriate intervention.

CONCLUSIONS

It is essential to enhance existing coordination and collaboration among the agencies in the federal government, and especially among those that are involved in risk assessment research. Clearly, that is something that the current Administration is emphasizing. We need to create more partnerships between government, both state and federal, industry and public interest groups. This includes labor unions, environmental groups, and anyone who has an interest in environmental health science research.

Another major issue is the setting of priorities in order to use available funding and other resources to our best advantage. According to OTA reports, only about $600 million is allocated to health science research, but billions of dollars are available for remediation and cleanup. In order to use the limited funding allocated to risk assessment research, we need to leverage that money with state agencies and industry. We need to set priorities to identify the most urgent environmental problems facing the American public. Once there is some agreement on those issues, we need to begin to use the available resources to address these concerns. If we do that, some of the contentiousness in this area will disappear, and we can begin to protect the health of the American people as Congress intended when NIEHS was established 25 years ago.

ETHICAL AND LEGAL
CONSIDERATIONS

Biomarkers and Occupational Health: Progress and Perspectives. 1995. Pp. 37-47

Legal Concerns in Worker Notification and the Use of Biomarkers in Medical Surveillance

Mark A. Rothstein

In the past few years, the United States has witnessed a remarkable and rapid transition from nuclear weapons production to environmental cleanup. With this transition, a number of new health and legal challenges have emerged. This paper puts some of these challenges in perspective by focussing on the legal requirements for worker monitoring and notification within the Department of Energy (DOE), which is responsible for the cleanup of the United States nuclear weapons sites.

A recent law, Public Law No. 102-484 (1992) mandates the notification of current and former DOE and DOE-contractor employees who are now being or who have previously been exposed to "significant health risks". Although this law provides that participation in any associated medical monitoring program is voluntary and that the information is to remain confidential, numerous ethical and legal questions remain unresolved.

Some of the legal questions to be addressed include the following: What is a "significant health risk"? Should the same standard apply to current and former employees? What disclosures need to be made to the individuals in order to obtain informed consent? What measures should be taken to ensure confidentiality? What laws regulating discrimination, privacy, and confidentiality are implicated by this program?

Complicating the already-difficult problems associated with notification and medical monitoring is the prospect of the use of biomarkers. The additional concerns raised by biomarkers include scientific questions (e.g., When are specific biomarkers sufficiently reliable and predictive to use?); ethical questions (e.g., Are there ethical differences between using these tests and other medical tests?); and legal questions (e.g., How does the use of biomarkers affect compensation claims?).

This discussion will address three important issues: first, the issue of when to notify at-risk employees and former employees; second,

the key elements of the notification process; and third, the legal consequences that could attach to notification.

WHEN TO NOTIFY AT-RISK EMPLOYEES AND
FORMER EMPLOYEES

First, with regard to Public Law 102-484 which was passed in 1992,
section 3162(a) indicates that the program of medical surveillance is
required to focus on both present and former DOE employees. Section
3163(2) indicates that DOE employees include contractor employees.
Thus, the potential magnitude of this program is quite startling. It
covers four categories of employees: current DOE employees, past DOE
employees, current contractor employees, and past contractor employees. These distinctions will be important in relation to the legal issues
and the legal rights that apply to each category.

The issue of when to notify individuals about certain health risks
is not unique in this area. It is a common theme in bioethics and health
policy. For example, it is an issue of great concern in the context of
the Human Genome Project (Macklin, 1992). When should an individual who has been identified as a carrier of a genetic disorder notify
siblings, children, parents, and so on? Similarly, when should a health
professional engage in such notification?

At one level, this issue may be viewed as a conflict between non-
maleficence and autonomy, two ethical principles. The principle of non-
maleficence could be invoked to curtail or eliminate employee notification. For example, some could argue that there is no point in notifying
people at risk of illness if the risks of exposure are well known, such as in
cases of exposure to radiation; or if there is no treatment available, such
as with some current late-onset genetic disorders; or if the individuals
are unlikely to understand the nature of the information that would be
given to them; or if they will suffer needless anxiety as a result of being
notified.

It could be argued that needlessly informing individuals would violate the first principle of medical ethics, *primum non nocere*, first do no
harm. On the other hand, such an approach might be considered overly
paternalistic and that at-risk individuals have a right to the information
regardless of the information's immediate utility.

There are practical arguments, as well, regarding the issue of when
individuals should be notified. Excessive notification requirements could

be viewed as extremely burdensome — in this case, extremely burdensome on DOE. It could also be argued that personnel and resources spent on notification might be better used in providing health care and environmental cleanup.

A final argument that could be used against disclosure of information is the fear that it will lead to litigation. This fear ultimately led to the defeat of the High-Risk Worker Notification Bill when it was proposed before Congress in 1987 (S. 79, H.R. 162, 100th Cong., 1st Sess.).

Fortunately, the policy debate surrounding whether to notify individuals at risk was resolved by Congress when it enacted P.L. 102-484. The public policy now is clearly established in favor of notifying such employees, although the elements of when and how to notify and whom to notify still have not been determined.

The Notification Process

The next issue is the key element in the notification process under P.L. 102-484. Section 3162(a) provides for the identification and ongoing medical evaluation of employees who are exposed to "significant health risks." The term "significant health risks" is not further defined in the statute. However, section 3162(b)(2)(B) refers to significant health risks under "federal and state occupational health and safety standards." Thus, one interpretation could be that only individuals who are exposed to hazardous or radioactive substances in excess of OSHA standards need to be notified and monitored. This would be an erroneous interpretation, at least in regard to current employees.

In 1980, the Supreme Court in *Industrial Union Department, AFL-CIO v. American Petroleum Institute*, 448 U.S. 607, while relying on the concept of significant risk, specifically stated that the requirement that a significant risk be identified is not a mathematical straightjacket. The agency is free to use conservative assumptions in interpreting the data with respect to carcinogens, so long as these assumptions are supported by a body of reputable scientific thought. It is preferred to risk error on the side of over-protection rather than under-protection.

In addition, the Supreme Court expressly sanctioned the use of the "action level" concept, where employers would be required to monitor the health consequences of employees even when those employees had exposures below those considered to be a safe level, or below the OSHA standard. The purpose of the action level concept was simply to recognize the wide degree of human variability in response to environmental agents.

Thus, a program that is based on the principle of information sharing and preventive medical surveillance should be broadly construed. And there is ample legal precedent to do so.

With regard to former employees, these individuals should probably not be informed of prior exposures and possible consequences along the identical lines as those of current employees. Clearly, if a program were adopted of notifying every single former employee who was exposed to any hazardous substance, it would be a mammoth and meaningless task, with little compliance expected from employees. Former employees should be notified only when (1) the exposures were more than *de minimis*, although not necessarily in excess of applicable standards; or (2) there is scientific evidence that exposure at that level could result in adverse health effects; or (3) where medical surveillance of similarly exposed individuals has been shown to have some value.

The other key provisions of P.L. 102-484 are as follows: Evaluations and tests should be made available to all employees identified as being at risk. Employee participation in the program is voluntary. Notifications of test results must be provided in a form that is readily understandable by the employee. Finally, employee privacy must be maintained with respect to all medical information that exists in a personally-identifiable form.

These four elements, which are expressly written into the statute, are important for a variety of reasons. Obviously, these are extremely important concerns from an ethical standpoint, but they are also extremely important from a legal standpoint. For example, if the program of medical surveillance were not voluntary and confidential, then there may well be legal issues raised by the imposition of such a program.

The fact that employee participation in the medical monitoring program is voluntary and confidential also takes some degree of pressure off the science. If it were otherwise, if the program were involuntary, and the results were widely disseminated, there would statutory and, indeed, even constitutional problems with using newly developed scientific techniques such as biomarkers.

The possibility exists that there could be legal liability on the part of DOE if there were improper performance of these statutory duties. For example, negligence in diagnosis or treatment or negligently supplying medical information could result in claims of negligence or medical malpractice against the Department of Energy under the Federal Tort

Claims Act (28 U.S.C. §§2671-2680). In addition, because the program must be voluntary, certainly at least for former employees, it must be clear that informed consent must be obtained from any person who agrees to receive information and to undergo medical evaluation.

On the issue of informed consent, many of the premises underlying informed consent have not been realized in the clinical setting. What started out as a concept designed to empower patients, has actually turned into a practice that merely protects physicians. Too often, obtaining informed consent has become a *pro forma* exercise (Katz, 1977). With regard to the DOE program, it is important to carefully study the issue of informed consent, the consequences of participation in the DOE program, and the elements that need to be included in a program of informed consent.

LEGAL CONSEQUENCES OF NOTIFICATION

The legal concerns related to the notification process fall into three general categories: the possibility of employment discrimination, medical records access and confidentiality, and the effect of notification on compensation.

Discrimination

First, with regard to the discrimination issue, the primary concern is that the identification of certain at-risk current employees, or even former employees, will lead to subsequent discrimination in employment practices: This is where it is necessary to separate the different categories of employees.

Contractor employees are covered by the Americans with Disabilities Act (ADA) of 1990 (42 U.S.C. §§12201-12213). This law applies to all private sector employers with 15 or more employees. It prohibits discriminatory practices in hiring, work assignment, promotion, wages, hours, benefits, layoffs, discharge, or any other term or condition of employment. It prohibits this discrimination on the basis of physical or mental disability.

Unfortunately, it is not yet clear whether an individual who is presymptomatic for a genetic disorder or who is presymptomatic for or otherwise measured to have subclinical effects of exposures is covered under the Americans with Disabilities Act (Rothstein, 1992). This is an issue that deeply concerns those who are involved with the Human Genome Project.

There are two other provisions of the ADA that are of particular relevance. The first one deals with the medical examination provisions under the Americans with Disabilities Act and the second one deals with what constitutes a direct threat under the statute.

The medical examination provisions of the ADA are contained in section 102(d). It essentially splits medical examinations into three parts depending on when they are performed. Section 102(d)(2) provides that it is unlawful for an employer covered by the ADA to conduct pre-employment medical examinations or to ask any individual whether he or she has any disability. At the pre-employment stage, before an offer is made, the employer is permitted to ask only if the individual, with or without reasonable accommodation, can perform the essential functions of the job.

After a conditional employment offer is made, section 102(d)(3) applies. This provides that the employer may conduct a preplacement examination, which the statute refers to as an "employment entrance examination," prior to assigning the individual to work. According to the Equal Employment Opportunity Commission (EEOC), which enforces the ADA, these examinations need not be job-related and may be comprehensive and mandatory (29 C.F.R. §1630.14[b]). This is another area of great concern because it facilitates the collection of sensitive but unusable medical records as well as providing the information that could be used for a discriminatory purpose.

It is important to note that the laws in 11 states, which are not preempted by the federal law, limit all medical examinations to job-related matters (Alaska Stat., etc.). Under the ADA, an individual's conditional offer of employment may not be withdrawn on the basis of a subsequent medical examination unless the medical criteria used are job-related and consistent with business necessity. Therefore, the anomaly exists that, under the ADA as currently interpreted by the EEOC, employers are permitted to require individuals to submit to medical examinations whose results may not be used.

The burden of justifying the exclusion of an individual on the basis of a medical examination is on the employer. Additionally, if an employer conducts such examinations, they must be conducted on all individuals seeking employment in a certain job category. And all information obtained must be collected and maintained on separate forms and stored in separate files. The only exceptions are, first, supervisors and managers may be informed regarding necessary restrictions on the duties of the individual employees and any accommodations that must be made to permit them to perform their work. Second, first aid and safety personnel may be informed, when appropriate, if the disability might require first aid treatment or emergency care. Third, government officials monitoring compliance with the ADA are entitled to request relevant information.

Section 102(d)(4) of the ADA covers medical examinations of current employees. This section provides that all medical examinations or inquiries of current employees must be either job-related and consistent with business necessity, or voluntary. This appears to be consistent with P.L. 102-484 in that all medical examinations of these current employees are on a voluntary basis. However, because medical surveillance of the effects of workplace exposures would be considered to be job-related, a detailed program consistent with P.L. 102-484 would not violate the ADA.

The other provision of the ADA that is relevant to this discussion is section 103(b). This provision states that the term "qualification standards" may include a requirement that an individual not pose a direct threat to the health or safety of other individuals in the workplace. This section has been broadly construed by the EEOC, unlike some of the other sections of the Act. The EEOC has issued an interpretive regulation which construes this term much more broadly than the statute. It provides that the term "direct threat" means a "significant risk of substantial harm to the health or safety of the individual or others that cannot be eliminated or reduced by reasonable accommodation" (29 C.F.R. §1630.2[r]). This means that if medical surveillance of a current employee indicates the adverse effects of prior exposure, that employee may not be discharged or reassigned against his or her wishes so long as that individual is capable of performing the essential functions of the job with or without reasonable accommodation. And the individual may be excluded from employment only if his or her employment constitutes a direct threat to the health or safety of that individual or others.

In the case of an employee whose medical monitoring demonstrated that the person was at increased risk of future illness or suffering from the subclinical effects or even minor symptoms resulting from prior exposure, would that justify removing the individual in an effort to protect his or her health? This would probably not justify removing the individual. This exception has been construed in an extremely narrow way. The employer has to demonstrate a direct and immediate threat to the health and safety of the employee. Nevertheless, the employee should be informed of the potential consequences of continuing the exposure that has been detected by the biological monitoring. But, whether that person remains in that work setting seems to be a matter for that individual, and perhaps with the input of his or her own physician, to decide.

The Supreme Court case of *International Union, UAW v. Johnson Controls, Inc.*, 499 U.S. 187 (1991), specifically stated that it was unlawful for an employer to remove all fertile women from exposure to inorganic lead in the workplace in order to protect potential hypothetical fetuses. The Supreme Court unanimously held that the decision whether to accept such risks of workplace exposure was specifically and exclusively within the domain of the woman.

For Department of Energy employees, it must be reiterated that the Americans with Disabilities Act specifically does not apply to federal government employees. Nevertheless, section 501 of the Rehabilitation Act does (29 U.S.C. §791). Moreover, Congress has indicated that the same principles used under the ADA should apply to section 501 of the Rehabilitation Act.

Confidentiality

With regard to the confidentiality of medical information, it depends on the category of employees. For DOE employees and former employees, these individuals are covered by the Privacy Act (5 U.S.C. §552a), which prohibits any federal agency, including DOE, from disclosing its records on any employee, including medical records, without the express consent of that individual. In addition, the agency must grant that employee a right-of-access to his or her own records and permit the copying of any record in the file. For contractor employees and former employees, the Privacy Act does not apply because DOE does not maintain the records. In addition, OSHA's access to exposure and medical records regulation (29 C.F.R. §1910.20) does not apply because section (4)(b)(1) of OSHA exempts DOE contractor employees (29 U.S.C. §653[b][1]); however, there are currently proposals in Congress that would change that.

The remedies available to contractor employees for wrongful disclosure of medical records come from two sources. First, section 102(d)(3) of the Americans with Disabilities Act specifically mandates that employer-maintained medical records must be confidential. Second, five state laws grant employees a right of access to some or all of their medical records within their employer's possession (Iowa Code Ann., etc.). The remedy for wrongful disclosure of this information under state law usually includes a common law action for invasion of privacy.

Compensation

The final issue is that of compensation. There are two broad categories. First is workers' compensation. Current employees who sustain occupational injury or illness may file a claim for medical benefits and wage loss under workers' compensation laws. For DOE employees, this is pursuant to the Federal Employees' Compensation Act (5 U.S.C. §§8101-8193). For contractor employees, this would be pursuant to state workers' compensation laws. It is important to note, however, that the chances of an individual employee receiving compensation for occupational illness remain small. The employee has the burden of showing that the illness from which he or she suffers is not an ordinary disease of life. While only 10 percent of workplace injury cases are contested by employers, 63 percent of occupational disease claims are contested by employers (Darling-Hammond and Kniesner, 1980). In fact, most instances of occupational illness do not result in the filing of a claim for workers' compensation, primarily because of the problem in diagnosing occupational diseases (Schroeder and Shapiro, 1984).

For former employees, both DOE and contractor employees, if they discover that their illness was caused by former workplace exposures, in most cases the statute of limitations period under which they have to file a claim will not begin to run until they discover the work-relatedness of their condition. Therefore, they may still file a claim even if their last exposure and their last work for DOE or a contractor was many years before. In some jurisdictions, however, the statute of limitations period begins to run with the "last injurious exposure."

With regard to the issue of common law actions for damages, workers' compensation laws in every state bar the filing of lawsuits for damages resulting from injury or illness in the workplace. There are a number of exceptions to this rule, however. One potentially relevant exception is based upon the concept of fraudulent concealment of medical information and, to a lesser extent, fraudulent concealment of hazards from the employees. For example, if the employer knew through medical surveillance that certain individuals were showing early symptoms of occupational disease or maybe even advanced symptoms of occupational disease and did not inform the individuals, then there may be liability above and beyond workers' compensation for the aggravation of the condition caused by the failure to obtain medical care in a timely manner (Santiago v. Firestone Tire & Rubber Co., etc.). Biomonitoring also may be valuable at some point in establishing the crucial element of causation that is needed for both workers' compensation claims and potential private litigation.

CONCLUSIONS

This discussion has briefly covered a variety of legal issues that could be raised by this program. Undoubtedly there are many more. The framework seems to be already in place, and DOE is to be applauded for its efforts in trying to determine how to legally, ethically, and efficiently implement this far-reaching statute. The manner in which DOE does this may well set a model for the rest of the country for private sector employers who are also concerned about these issues.

REFERENCES

Alaska Stat. §18.80.220(a)(1); Cal. Gov't Code §12940(d); Colo. Rev. Stat. §24-34-402(1)(d); Kan. Stat. Ann. §44-1009(a)(3); Mich. Comp. Laws Ann. §37.1206(2); Minn. Stat. Ann. §363.02(1)(8)(i); Mo. Ann. Stat. §213.055.1(1) (3); Ohio Rev. Code Ann. §4112.02(E)(1); 43 Pa. Cons. Stat. Ann. §995(b) (1); R.I. Gen. Laws §28-5-7(4)(A); Utah Code Ann. §34-35-6(1)(d).

Darling-Hammond, L., and T. Kniesner. 1980. The Law and Economics of Workers' Compensation. Santa Monica, California: Rand Corporation.

Iowa Code Ann. §§91B.1, 730.5; Me. Rev. Stat. Ann. §631; Ohio Rev. Code Ann. §4113.23; 40 Okla. Stat. Ann. §191; Wis. Stat. Ann. §103.13(5).

Katz, J. 1977. Informed Consent–A Fairy Tale? University of Pittsburgh Law Review 39:137.

Macklin, R. 1992. Privacy and Control of Genetic Information. Chapter 9 in Gene Mapping: Using Law and Ethics as Guides, G.J. Annas and S. Elias, eds. New York: Oxford University Press.

Rothstein, M. A. 1992. Genetic Discrimination in Employment and the Americans with Disabilities Act. Houston Law Review 29:23.

Santiago v. Firestone Tire & Rubber Co., 274 Cal. Rptr. 576 (Ct. App. 1990) (leukemia); Delamotte v. Unitcast Div. of Midland Ross Corp., 411 N.E.2d 814 (Ohio App. 1978) (silicosis) (sample cases).

Schroeder, E. P., and S. A. Shapiro. 1984. Responses to Occupational Disease: The Role of Markets, Regulation, and Information. Georgetown Law Journal 72:1231.

Biomarkers and Occupational Health: Progress and Perspectives. 1995. Pp. 48-51

Biomedical Research Ethics Related to Biomarkers

William N. Rom

There are high stakes in biomedical research, with money, prestige, and important publications as the potential rewards. For those on front stage, ethics has played second fiddle in recent times. For good reason, the Department of Health and Human Services and the U.S. Congress have grappled with the issue of biomedical research integrity. Biomedical research ethics covers illegal acts, unethical conduct, and reprehensible behavior. Illegal acts include outright fraud, breach of contract, and theft that violates federal, state, or other laws. Unethical conduct includes plagiarism, fabrication, and falsification. Reprehensible behavior includes such acts as suppression of data, lack of attribution, and conflict of interest. The basis for a lack of ethics usually involves greed, arrogance, and ambition, and their existence can stem from a lack of supervision by higher levels of authority or appropriation of power by authority without challenge by members of the organization.

The three case studies presented below provide examples of how ethical considerations can and should become an integral part of scientific and medical activities.

Case Study 1: Biomarkers and Beryllium

Biomarkers can be used to measure exposure or response to exposure and may be considered a measure of disease or risk for developing disease. Few prospective studies have been conducted to determine the rate of disease in those individuals who are biomarker-positive or biomarker-negative. Also, the relationship of removal from exposure and the persistence of biomarkers in an individual's tissues after that exposure ceases is poorly understood.

In a study of beryllium processors handling bertrandite ore in Utah, we evaluated the blood lymphocyte proliferation (transformation) test as a biomarker of disease resulting from exposure (Rom et al., 1983). The company paid the laboratory directly for the test results. Each worker signed an informed consent form and then was administered a questionnaire, followed by spirometry, chest radiography, and blood drawing. For these workers, 15.9 percent (13/82) of the test results (according to the method of Deodhar et al., 1973) were positive. However, there were no abnormalities consistent with beryllium disease on chest radiographs or lung function tests.

A major component within the concept of informed consent is the notification of workers. This company rejected the idea of notifying the workers about the results of the biomarker test because it was experimental and because no one had clinical signs of berylliosis. We rejected this notion outright and notified the workers about their test results, including the biomarker test, and presented them with a simple explanation of the test and what we thought it meant. We were particularly concerned about whether this test indicated exposure since none of the workers demonstrated physiologic evidence of disease.

A cardinal rule in clinical medicine is to repeat any test with abnormal results. The company refused us further access to the plant to repeat the lymphocyte proliferation tests to beryllium. We negotiated with the union who cooperated with us in conducting a follow-up study 3 years later at the community hospital. Immediately after our first study, the furnace operation had been discontinued and the bagging operation, where exposure was notably difficult to control, was enclosed and automated. As a result, exposures were significantly reduced. Interestingly, only 5/61 workers (8.2%) were now positive, and 8 of 11 previously positive were now negative. The workers with positive beryllium lymphocytic proliferation blood tests who had been exposed in these operations had reverted to negative several years later. We proposed that this test may be used as an indicator of exposure as well as disease.

Case Study 2: Biomarkers and Asbestos

While at the Pulmonary Branch, National Heart, Lung, and Blood Institute, National Institutes of Health (NIH), I conducted a study to evaluate the fundamental mechanisms of asbestosis using bronchoalveolar lavage (Rom et al., 1987). We measured the mediators released by alveolar macrophages that could be considered biomarkers: fibronectin and alveolar macrophage-derived growth factor which stimulate fibroblast replication and enhance matrix accumulation, and superoxide anion (O_2-) and hydrogen peroxide (H_2O_2) which injure alveolar epithelial cells. Mediator release from the alveolar macrophages of asbestos workers was significantly greater than that released from controls' alveolar macrophages.

Several ethical dilemmas arose in relation to this study. First, these workers had been exposed to asbestos while insulating buildings at NIH and other U.S. government facilities with asbestos, and they were now being studied at NIH. All of these workers had lawsuits aimed at the suppliers of the asbestos (tort liability or "third party" lawsuits). This irony was superseded by the fact that the major defense expert was a government scientist at the Pulmonary Branch, although no testimony was presented regarding any of our study subjects. Moreover, if any of the manuscripts resulting from this study dealt with asbestos, there would be a delay of more than two years prior to submission for publication. In addition, if the data were disadvantageous to the position of the asbestos industry, approval authorization for publication would be elevated all the way to the Acting Director of NIH. It was found that there were large amounts of crystalline asbestos in the bronchoalveolar lavage cells of these asbestosis patients, and the asbestos industry alleged that this type of asbestos was less harmful than other types and would break down over time (more than 20 years).

It is self-evident that the issue of government scientists providing consultation to the industry being studied is an extraordinary conflict of interest and should be proscribed, including the use of personal time by any government scientist.

Case Study 3: Conflict of Interest with Biotechnology Companies

Biomarkers and genetic tests are patentable ideas and discoveries. In order to market these ideas, a thriving biotechnology industry has emerged. Venture capital has lead to the founding of dozens of firms, yet the product lines promoted by these companies remain limited. This industry provides much of the funding that supports university-based research efforts in this technical area. In addition, university scientists participate as consultants in these companies and work together to commercialize products.

An example of reprehensible conduct recently occurred in relation to the product of collaboration among several university scientists. A patent application was submitted for this idea by a biotechnology firm associated with one of the scientists, but the other collaborating scientists and the university were excluded. Conflict of interest in industrial consultations needs to be better defined.

Remedies

Ethical conduct and integrity need to be given high priority in biomedical research programs. Principles of ethics in research should be taught to medical and postgraduate bioscience students and to research fellows. University administrators need to take research ethics issues seriously. As a prerequisite to receiving NIH or U.S. government grants, recipients should demonstrate that their institutions use a systematic approach to evaluating potential ethics issues. This might include a research ethics council that adjudicates charges of violation of research ethics and immoral behavior in addition to scientific misconduct. Also, there should be a mechanism of enforcement that universities (and the U.S. government research laboratories) utilize, e.g., penalties that range from docking a month's salary for lesser infractions to a more serious censure for more serious infractions. Particularly in an area like biomarker development and use, it is essential that the scientific community take an aggressive stance on maintaining research integrity.

REFERENCES

Deodhar, S. D., B. Barna, and H. S. Van Ordstrand. 1973. A study of the immunologic aspects of chronic berylliosis. Chest 63:309-313.

Rom, W. N., P. B. Bitterman, S. I. Rennard, A. Cantin, and R. G. Crystal. 1987. Characterization of the lower respiratory tract inflammation of nonsmoking individuals with interstitial lung disease associated with chronic inhalation of inorganic dusts. American Reviews of Respiratory Diseases 136:1429-1434.

Rom, W. N., J. E. Lockey, K. M. Bang, C. Dewitt, and R. E. Johns. 1983. Reversible beryllium sensitization in a prospective study of beryllium workers. Archives of Environmental Health 38:302-307.

Biomarkers and Occupational Health: Progress and Perspectives. 1995. Pp. 52-57

Biomarkers: The Down Side

Lewis Maltby

Nowhere are the benefits and risks of scientific progress more evident than in the use of biomarkers. On the one hand, the use of biomarkers has enormous potential to spare people from the ravages of industrial disease. Workers in United States Department of Energy (DOE) facilities are exposed to many chemicals that are potentially hazardous to their health. These include beryllium, chrome, nickel, asbestos, carbon tetrachloride, and formaldehyde. If there are biomarkers that can tell us when a given worker's exposure is approaching the point of danger so that this worker can be removed from the hazardous area, using them could protect workers' health or even save their lives.

But biomarkers are only a tool, and like all tools, they can be misused. One way biomarkers can be misused is for managers to use the information that biomarkers provide to make decisions that should be made by the workers themselves. The decision to remove a worker from a hazardous section of the facility when their exposure reaches a dangerous level may seem an obvious choice that would be welcomed by the worker, but this is not necessarily true. Few workers would object to a temporary transfer that increased their safety if it were without any consequences. But what if the transfer results in a reduction in pay? There may not be a position of equal pay into which the worker can transfer. This may not specifically be a concern at DOE facilities where there is a system of pay protection for transferred workers but there remains the concern of a loss of advancement opportunities and the inability to get the same overtime opportunities.

In many cases, the length of time until the biomarker levels indicate that the worker can safely return to his or her original job will be short, and the employer will be willing to continue the worker's original pay rate during the transfer. But if the levels of biomarkers never return to normal or require a protracted transfer, some workers may be forced to choose between maintaining the ability to support their families and having a safe job.

No one should ever be faced with such a choice. But, if such a choice must be made, the workers should at least make it for themselves. If biomarkers become a tool for well-meaning but naive managers to take life and death decisions away from employees, they will only have made a bad situation worse.

The use of biomarkers also sows the seeds for employment discrimination. What happens when an employee who has had a temporary biomarker-related transfer attempts to seek employment outside of the facility? Future prospective employers might suspect that this person will require such inconvenient transfers again. Or, they might feel that the employee was a general health risk and is somehow tainted because of his or her previous experience. The cost of reducing an employee's risk today through the use of biomarker technology may be a lifetime of employment discrimination.

Some may question whether employers would really engage in discrimination over a factor of such questionable medical validity and dubious economic impact. But based on similarly sketchy evidence of economic impact, literally thousands of employers discriminate against potential employees who smoke or drink off-duty, have risky hobbies, or are fat.

Of course, in order to engage in such discrimination, future employers need to have access to information about the person's previous medical and employment experience. Obtaining access to that information is alarmingly easy. The employer needs only to require an applicant to consent to the release of all previous medical and employment records as a condition for the consideration of their application. An applicant is not legally required to give the consent; however, they will have little choice if they want the job.

An employer might even be able to get this information without the applicant's knowledge. A Louisiana company will sell the complete workers compensation records for any of the one million people in their data bank to a prospective employer or anyone else who asks. Thousands of people who have suffered injuries on the job have been victimized twice because they dared to apply for workers compensation. Most of them never realized the reason why. Employers have no legal obligation to tell people why they were turned down for a job. If this can happen with workers compensation records, it might well happen with biomarker records.

In theory, the Privacy Act should prevent federal records from becoming an open book. The Privacy Act requires that federal records about individuals be released only for purposes that are consistent with the purpose for which the data were originally collected. But the General Accounting Office found that 56% of federal agencies routinely release records with no analysis of whether the disclosure is legal. In most cases, the agencies do not even ask the purpose for which the data will be used. In addition, federal agencies routinely share data among themselves. Therefore, even if an agency like DOE is scrupulously careful in its handling of sensitive records, their confidentiality can be lost when these records are shared with agencies that are less careful.

To make matters worse, there is no effective legal protection against such discrimination. The Americans with Disabilities Act (ADA) appears to offer protection, but close examination reveals this to be an illusion. To begin with, not all physical conditions are covered by the ADA. To qualify as a disability under federal law, a condition must "substantially limit an individual in a major life activity." This means that a condition must be fairly serious before it is covered by the ADA. A condition that is painful and expensive to treat but does not interfere with a person's ability to carry out the normal activities of life is not covered by ADA. Most conditions indicated by biomarkers will meet this test, but some may not.

Most of the people affected by a biomarkers program, however, will not have the disease in question. They have a physical marker that indicates that they may be at risk for developing the disease. The case law on such potential disabilities is good. Most knowledgeable observers, for example, believe that people with genetic conditions that will lead to future disease are covered. Whether the law will extend to people whose biomarkers indicate a high level of exposure to a toxic substance remains to be seen. The biggest stumbling block may be the absence of any abnormal physical condition at the time the discrimination occurs. Unlike the person with a bad back or the gene for Huntington's disease, the person affected by a biomarker program may be perfectly normal at the time they suffer employment discrimination.

For people whose biomarker levels indicate a high level of exposure to a toxic substance, the greatest problem with the ADA stems from a generic weakness in the act. The ADA permits employers to require a medical examination, including a medical history, once a conditional offer of employment has been made. This examination and history do not have to be limited to conditions that are related to the job in question. An employer is free to look for conditions that would be illegal to consider in the employment decision. Since employers are not required to disclose their reasons for rejecting an applicant, there is an enormous potential for abuse. While this problem is not unique to people with elevated levels of biomarkers, they may be particularly at risk since their conditions, unlike those of many people with disabilities, will first become known to the employer through the medical examination.

Beyond the world of employment, there are general privacy concerns raised by biomarker programs. These programs inherently involve the collection of sensitive medical information about individuals. If this information were to be released without the person's consent, this would be a serious loss of privacy. As discussed above, the track record of federal agencies in handling sensitive records is far from impressive.

These concerns may not be sufficiently serious to reject the idea of a biomarker program out of hand. But they do argue for restricting biomarker programs to situations where they can provide a clear-cut indication that a worker is at risk for a serious medical condition. The risks of future job discrimination and loss of privacy may be worth running in a life and death situation, but this is not true for every medical condition.

These considerations also argue that participation in a biomarkers program ought to be voluntary. While participation clearly has benefits, it is equally clear that these benefits come at a cost. The relative value of these benefits and costs is a highly personal matter. None of us is wise enough to make this decision for another person, and even if we were, none of us has the right. The decision of whether to participate in the program should be left to the individual worker who must live with the consequences of the decision.

Finally, the records of any biomarkers program must be kept confidential, far more so than is generally the case.

Fortunately, Public Law 102-484, section 3162 which would provide the statutory basis for the DOE biomarkers program, addresses all of these concerns. It provides that the program will cover only those workers with a "significant health risk." It provides that participation in the program must be voluntary, and it requires that "privacy is maintained."

But these statements are all broad generalizations, and as Mies Van Der Rohe once said, God is in the details. For example, what is a significant health risk? Someone who wants the broadest possible biomarkers program could argue that any condition that requires medical treatment is significant. This is not the most reasonable interpretation, but a court might accept it.

And what does it mean in practice that participation in the biomarkers program be voluntary? Is it voluntary if participation is a condition of employment? This is undoubtedly not what Congress meant, but some courts have taken the position that an agreement which someone makes in order to keep his or her job is voluntary. What if there is no overt official pressure, but the worker's immediate supervisor makes it known that he or she will be displeased if the worker does not "volunteer"?

And what does it mean in practice for privacy to be maintained? Does it mean that information is released only for a legitimate reason? By that standard, virtually any disclosure could be defended. Even if it means no disclosure without the written consent of the worker, there is reason for concern. An employee who is asked for such "consent" as part of the application process for a subsequent job will have little choice but to comply. The same will be true regarding applications for insurance.

Finally, there is the problem of how these standards will be enforced. Without effective enforcement mechanisms, these rules will be meaningless. But the statute does not specify how they will be enforced. This is particularly alarming in light of DOE's decision to store certain sensitive medical data in other countries, such as Italy. What recourse does an American worker have if an Italian data bank improperly releases information about him or her? Does Italian law provide a remedy? And, even if it does, how can the average worker enforce his or her rights in a court and country that would be very costly to reach?

Answers to these detailed questions will have to be provided by the regulations that are still being formulated. While these specific questions are reflections of larger issues to which there are no easy answers, we do have some approaches to suggest.

The best way to ensure that participation in a biomarkers program is truly voluntary is to set the program up in such a manner that those with an incentive to apply pressure do not know who has enrolled. The program should be run by a separate team of health care professionals completely removed from line management.

Confidentiality can best be maintained by allowing requests for disclosure only when the employee can refuse disclosure without penalty. Disclosure, even with written consent, should not be allowed when it is part of a contract of adhesion. For enforcement to work, the individual worker or his or her representative must have the right to initiate a complaint and participate fully in its resolution. Leaving enforcement exclusively in the hands of a government agency whose officials are not directly affected by the issue, rather than to the workers, is inadequate.

The Department of Energy has made a good start in attempting to provide the benefits of biomarkers to its employees. But several important issues need to be resolved before this goal is completely achieved.

PRIORITIES, COSTS, AND STANDARDS

Biomarkers and Occupational Health: Progress and Perspectives. 1995. Pp. 61-69

Application of Biomarkers: Getting our Priorities Straight

Marvin S. Legator

The rapid development and advancement of techniques in the area of molecular biology will allow for parallel progress in the use of biological markers for monitoring human populations. These biological markers can be used to detect exposure to specific agents, characterize the effects of toxins, and determine susceptibility to specific chemicals. For several years, it has been possible to determine adduct formation, cytogenetic effects and frequencies of somatic cell mutations in high-risk populations exposed to carcinogenic agents. When correctly applied and interpreted, these techniques have played a major role in limiting exposure to hazardous substances such as benzene and ethylene oxide. However, existing technologies have been underutilized.

This paper discusses the need to prioritize efforts to apply biomarkers in the characterization of human exposure to toxic agents, gives examples of monitoring programs in industry, and presents an integrated study designed to determine the effects of low-level exposure to a widely used industrial chemical.

THE NEED FOR PRIORITIZATION

Limited Resources

For biological research in general, and especially research related to toxic substances, there is a limited amount of available funding. The National Institute for Environmental Health Sciences (NIEHS) budget for research on environmental pollutants has not substantially increased in recent years, and the Environmental Protection Agency (EPA) and the Agency for Toxic Substances and Disease Registry (ATSDR) continue to restrict funding in this area. These funding limitations mandate that every project should be carefully evaluated to ensure the best use of available funds. The application and development of biological markers should be carried out so as to maximize our ability to obtain useful information that can benefit populations at risk.

Information Gaps in the Evaluation of Environmental Toxins

A few years ago, a National Academy of Sciences-National Research Council (NAS-NRC) committee concluded that for more than 75% of industrial chemicals, there is either minimal or a complete absence of toxicological data (Nelson and Upton, 1987). Examples include methyl· isocyanante, which was released during the Bhopal disaster, a widely used gas additive, methyl tertiary butyl ether (MTBE), and butadiene which will be discussed later.

Need to Study Low-Level Chronic Exposures to Widely Used Chemicals

Exposures to chemicals, either in the workplace or in community settings (such as near hazardous waste disposal sites) usually occur at comparatively low levels over a prolonged period of time. Exposures to known carcinogens such as vinyl chloride, benzene, and butadiene are currently at levels far below what they were in the past. Current epidemiologic data on disease incidence reflect exposures that occurred many years ago as illustrated by the case of butadiene. Although there is very little available information on exposures in styrene-butadiene rubber and butadiene production plants during World War II, it is likely that the levels of butadiene to which workers were exposed during those years were much higher than those occurring today. The existing permissible exposure limit is 1000 ppm (Fajen et al., 1990); however, the Occupational Safety and Health Administration proposed in 1990 that it be lowered to 2 ppm, and many companies have reduced exposure levels to below 10 ppm as a result of studies that demonstrated the carcinogenic effects of butadiene in rodents (Melnick et al., 1990; Owen and Glaister, 1990).

Research needs should be prioritized in order to conserve available resources. Biomarkers should be used to study chronic low level exposure to chemicals in instances where there is insufficient information to allow for hazard evaluation in human populations. The need to utilize existing biomarkers is imperative given the insensitivity of studies using chronic disease outcomes (conventional epidemiology) for assessing the effects of prolonged low-level exposure to chemicals (Neutra, 1990).

Studies of high-level exposure to agents for which there is already a considerable database should not be considered as a priority item.

Biomonitoring Programs in the Chemical Industry

In 1964, the Texas Division of the DOW Chemical Company initiated a cytogenetic testing program (Legator, 1987). By 1974, 43,044 shorthand and conventional karyotype tests had been performed on 1,689 workers involved in the production of various chemicals. During the same period of time, 25,104 karyotype tests were performed on 1,302 individuals prior to their employment (Kilian and Picciano, 1976). All prospective employees were given a pre-employment physical examination that included a cytogenetic analysis that would result in the accumulation of a large pool of control data and would also serve as a reference point for future clinical investigation of the individual.

Many of the problems that might be anticipated from such an industrial screening program were successfully obviated in the DOW facility. Some individuals were found to have chromosome abnormalities that were not associated with their work exposure, and these people were referred to suitable medical experts for genetic counseling. Because the program was carefully explained to the employees, little resistance was encountered, and the program seemed to be well accepted.

During the course of this program, there were many important improvements in cytogenetic monitoring. The minimal number of cells to be analyzed was determined; questionnaires to be used prior to cytogenetic study were developed; and a computer assisted program was designed. The DOW management was pleased with its own program and felt that DOW was protecting its employees with the latest state-of-the-art cytogenetic screening procedures. Even at that point in time, about 20 years ago, cytogenetic monitoring was considered to be a reliable and objective method of evaluating possible genetic changes.

Many companies within the chemical industry would probably be ideal candidates for such studies. They employ stable populations and subpopulations of significant size that are involved at least to some extent in the production of chemicals, some of which are suspected of being genetic hazards. In many cases, accurate records of continuously monitored exposure to specific compounds, alone or in combination, are available. The number of study subjects is generally large enough for good study design and statistical analysis. Employee turnover rates in this industry tend to be low, there is adequate technology for accurate monitoring, and there are scientific capabilities for evaluating most changes found in the individual. A substantial number of employees (clerical and administrative personnel, for example) have no contact with the substances to be studied, and data from these groups are suitable for matched control evaluation. Interdisciplinary study design is enhanced by the ready availability of findings from periodic medical examinations and records from comprehensive health insurance plans.

Studies of the chemical industry work force and its subgroups would be useful in verifying the appropriateness of existing standards and providing information about the need to revise other standards well before an overt hazard is imminent. Findings of this nature serve the general population as well as the employees by ensuring that appropriate protective measures can be taken upon early identification of genetic risk (Kilian and Picciano, 1976).

In 1977, the DOW Chemical Company in Texas evaluated workers exposed to epichlorohydrin and to benzene. Both studies showed an increased number of chromosomal breaks associated with low-level exposure. These results are not surprising, since benzene is a known carcinogen and epichlorohydrin had been shown to induce chromosome abnormalities in a previous study. Shortly after these findings, DOW's extensive screening program was phased out. However, the notable achievements of the program included the following:

- The development of a comprehensive medical questionnaire to help determine possible factors that could influence the cytogenetic outcome.

- The development of an educational program to help workers understand the procedures and their significance.

- Referrals for genetic counseling of workers with abnormal findings not associated with any chemical exposure.

- The accumulation of a large data base, including data showing seasonal and temporal variation.

- The confirmation that benzene and epicholohydrin induced cytogenetic abnormalities in employees.

From all available information, the DOW program of cytogenetic screening of workers was a highly successful endeavor and played a leading role in developing the techniques for monitoring workers by short-term methods.

A Butadiene Study as a Model for the Application of Biomarkers

An ongoing study with 1,3-butadiene illustrates the use of biological markers to address key questions related to the safety of present day exposure to this chemical (Ward et al., 1994; Legator et al., 1993). Epidemiologic evidence indicates that 1,3-butadiene is a human carcinogen based on past workplace exposure levels, which were probably higher than present day levels. Metabolic studies of 1,3-butadiene indicate that there may be quantitative differences among species in the metabolism of this chemical. Thus, the question arises as to the sensitivity of humans to the mutagenic and carcinogenic effects of 1,3-butadiene. A biomonitoring study was designed to answer that question and to determine the following:

- Can biological effects be detected at present levels of workplace exposure?

- What are the most sensitive human genetic monitoring assays?

- Can biological endpoints be correlated with urinary metabolites of butadiene?

To address these questions, a group of workers in a butadiene extraction plant where exposure levels are approximately 1-3 ppm is being studied (Ward et al., 1994). This comprehensive study includes the participation of several laboratories and the evaluation of a variety of endpoints.

The study population consists of non-smokers (confirmed by plasma cotinine determination). Three groups of subjects were selected for this study: (a) a group exposed to 1,3-butadiene in production areas of the plant, (b) a group working in the same plant but in areas where exposures to butadiene were lower, and (c) an outside, non-exposed group consisting of employees in the Department of Preventive Medicine and Community Health at the University of Texas Medical Branch (UTMB) in Galveston.

Peripheral blood lymphocytes obtained from the study subjects were used for the genetic assays. Samples of blood and urine were collected from a total of ten workers each in the high- and low-exposure areas of the plant and from nine non-exposed subjects at UTMB. Three subjects with jobs that require time in both high- and low-exposure areas were included in the high-exposure group. Frequencies of HPRT mutant lymphocytes (Vf) were determined by autoradiography in eight subjects from the high exposure group, five from the low-exposure group and six from the non-exposed group. All high-exposed subjects from whom evaluable cultures could be obtained were assayed. The weighted mean HPRT.Vf in the high-exposure group ($3.84[\pm0.70] \times 10^6$) was significantly elevated relative to the low-exposure ($1.16[\pm0.27] \times 10^6; p < 0.05$) and non-exposed ($1.03[\pm0.70] \times 10^6; p < 0.01$) groups. The latter two groups were not significantly different from each other.

Cytogenetic analyses were performed on the ten samples each from both the high- and low-exposure groups. The non-exposed group has not yet been evaluated. No significant differences in structural chromosome aberrations were observed between the two exposed groups. However, when lymphocytes were exposed *in vitro* to x-irradiation, significant differences in the induction of dicentric chromosomes were observed. In the high-exposure group, the mean dicentric frequency was 0.219 (±0.053) as compared with 0.172 (±0.033) in the low-exposure group ($p < 0.03$).

A single metabolite, 1,2-dihydroxy-4-(N-acetylcysteinyl-S)butane was detected in the urine of all subjects. The mean levels in the high-exposure group were significantly different than in the low-exposed group or the non-exposed group. The frequency of HPRT mutant lymphocytes was highly correlated with the concentration of the butadiene metabolite detected in the urine. When the square root of the HPRT Vf and the log of the metabolite concentration were used, the correlation was $r=0.85$.

The three endpoints (HPRT Vf, dicentric frequency following radiation challenge, and urinary metabolite concentration) were all significantly elevated in the high-exposure group as compared with the control group. The strongest correlation was between the Vf and urinary metabolite concentration, with a weaker correlation between the Vf and dicentric frequency. No correlation was found between any of the outcome variables and age or years worked in the plant. The consistency of the results obtained to date with these two different assays and the correlation with the concentration of the urinary metabolite lead to the following conclusions:

1) Humans are sensitive to genetic damage induced by butadiene exposure.

2) Since present day exposure to butadiene at 1-3 ppm causes genetic damage, consideration should be given to lowering the permissible exposure level to below the 1-3 ppm level.

3) In view of these findings, including the identification of genetic endpoints and their correlation with a butadiene urinary metabolite, the alleged resistance of humans to the genotoxic effects of butadiene (based solely on pharmacokinetic and metabolic studies in different species) should be reconsidered.

4) The present report and additional studies in progress with this group of butadiene-exposed workers demonstrate the feasibility of detecting low-level chronic exposure by currently available genetic monitoring techniques. In addition, these results illustrate the advantages of conducting multi-endpoint studies and correlating biological effects with the presence of metabolites of suspected genotoxic agents.

It is anticipated that genetic monitoring will be used more frequently as an alternative to conventional epidemiologic studies with their limited statistical power. Genetic monitoring techniques are especially well suited for determining risks due to genotoxic agents at present levels of exposure.

Although the initial phase of the butadiene study uses a minimal number of subjects, efforts are now in progress to extend the data from this earlier investigation. This study provides an excellent example of how biological monitoring can be effectively used to provide data on probable future adverse health effects in human populations exposed to a widely used chemical. This information, which may play a role in the regulation of this chemical, could not have been determined as rapidly and at such a reasonable cost by alternate approaches such as disease outcome studies, i.e., conventional epidemiology.

CONCLUSIONS

Currently available biological markers can be used to determine the biological effects of chemicals associated with high levels of exposure, and for which initial information on these substances is not available. The insensitivity of conventional epidemiologic studies and the fact that these studies reflect past exposure further emphasizes the need to use biological markers.

It may be useful to generate a list of chemicals associated with widespread exposure, such as that developed by the International Agency for Research on Cancer for Class 2a carcinogens (known carcinogens with limited human data). This list should possibly also include chemicals for which minimal to no data exist. The fuel additive MTBE would be an example of a recently introduced high-volume chemical for which minimal data are available, and for which our need to establish safety guidelines is imperative. A list of priority chemicals for which biological monitoring studies are needed should maximize available resources and ensure that meaningful information of public health significance will be forthcoming.

REFERENCES

Fajen, J. M., D. R. Roberts, L. J. Ungers, and E. R. Krishnan. 1990. Occupational exposure of workers to 1,3-butadiene. Environmental Health Perspectives 86:11-18.

Kilian, D. J., and D. Picciano. 1976. Cytogenetic surveillance of industrial populations. In A. Highlander, ed., Chemical Mutagens: Principles and Methods for Their Detection, Vol. 4. Plenum: New York.

Legator, M. S. 1987. The successful experiment that failed. Pp. 465-486 in: Scientific Controversies, H. Engelhardt and A.L. Caplan, eds. Cambridge University Press.

Legator, M. S., W. W. Au, M. Ammenheuser, and J. B. Ward, Jr. 1993. Elevated somatic cell mutant frequencies and altered DNA repair responses in non-smoking workers exposed to 1,3-butadiene. Butadiene and styrene assessment of health hazards. IARC Scientific Publications 127:253-262.

Melnick, R. L., J. Huff, B. J. Chou, and R. L. Miller. 1990. Carcinogenicity of 1,3-butadiene in C57B1/6xC3H F_1 mice at low exposure concentrations. Cancer Research 50:6592-6599.

Nelson, N., and A. C. Upton. 1987. Generation of toxicological data on chemicals in the U.S.A. Definitive Science Journal 37:85-97.

Neutra, R. R. 1990. Reviews and commentary; counterpoint from a cluster buster. American Journal of Epidemiology 132:1-8.

Owen, P. E., and J. R. Glaister. 1990. Inhalation toxicity and carcinogenicity of 1,3-butadiene in Sprague-Dawley rats. Environmental Health Perspectives 86:19-25.

Ward, J. B., Jr., M. M. Ammenheuser, W. E. Bechtold, E. B. Whorton, Jr., and M. S. Legator. 1994. HPRT mutant lymphocyte frequencies in workers at a 1,3-butadiene production plant. Environmental Health Perspectives (in press).

Biomarkers and Occupational Health: Progress and Perspectives. 1995. Pp. 70-88

Costs of Developing a Large-Scale Biomarker Program to Monitor Cleanup Workers

Richard J. Albertini

There have been great advances over the past several years in our ability to detect genotoxic exposures and effects in humans. The driving force behind these advances has been an increasing awareness of the potential health impact of environmental mutagens and carcinogens in terms of chronic disease. Now that there is a variety of biomarkers for human studies, most of which have been "validated" in the sense that they reflect what they are supposed to, we are faced with decisions regarding their use.

The use of biomarkers for human monitoring raises several questions. First, what are the goals of monitoring, and can they be realistically achieved? Second, what are the costs of monitoring, and how do these costs relate to the benefits? An aspect of benefits is the potential savings of monitoring, which requires knowledge of the costs of not monitoring. Another set of questions concerns the ethics of monitoring. These last questions will not be considered here, except to acknowledge that some are very difficult, especially when biomarkers of *susceptibility* are used. However, the knowledge gained from the use of such biomarkers may be of utmost importance in terms of the estimation of individual risks.

Biomarkers

The term "biomarker" encompasses a large number of biological endpoints. For mutagens/carcinogens, these endpoints have been "ordered" according to stages in the process of environmentally mediated disease, i.e., cancer (Committee on Biological Markers of the National Research Council, 1987). According to this scheme, the deleterious agent in question first must be present in the environment, then must enter the body and be metabolized (if necessary) to a genotoxic intermediate, then must react with macromolecules, including the "target"

70

DNA, then must induce some chromosomal — or DNA — damaging event such as a chromosome aberration, micronucleus or somatic mutation. All of these events may be measured in surrogate tissue using surrogate endpoints, e.g., general chromosome aberrations or mutations in reporter genes. These are all biomarkers. Chromosomal or DNA damaging events may also be measured, using molecular techniques, in oncogenes or tumor suppressor genes. These last endpoints merge with markers of early frank pathology, such as altered cell structure or function, metaplasia or dysplasia, early carcinoma *in situ* and, finally, cancer. Damaging events in cancer genes may be considered as molecular manifestations of disease rather than as biomarkers.

The ordering of biomarkers in this way is useful in that it defines the stage in "pre-pathogenesis" being assessed by the different endpoints. Also, the goals of biomonitoring become obvious. The overall goal, of course, is the prevention of genotoxic disease. However, at least two intermediate goals are evident. One is to identify which humans have been exposed to environmental genotoxins. The other is to determine if these exposures have *in vivo* genotoxic effects, and to do so well before there is evidence of genotoxic disease. Exposure by a susceptible individual plus effect are the requirements for disease.

It is useful to classify biomarkers according to kind. Some are markers of *exposure/dose*, others are markers of *effect*, while still others measure *susceptibility*. Biomarkers of effect record responses in individuals who have had previous exposure to a genotoxic agent, but exposure/dose biomarkers do not, necessarily, indicate effects. Theoretically, it would seem that, as one moves through the scheme outlined above, one moves from biomarkers of exposure to biomarkers of effect.

The detection of environmental chemicals or their metabolites in body fluids indicates exposure, as does the presence of protein adducts. DNA adducts demonstrate that the critical target of genotoxicity has been exposed, thus reflecting primarily exposure/dose, although repair, an effect of the exposure, may modify this endpoint's "readout." Sister chromatid exchanges (SCE) are primarily exposure biomarkers. Chromosome aberrations and micronuclei require "processing" by the host to develop a damaging event. When heritable over the somatic line, chromosome changes may result in cancer. Chromosome level changes are effect biomarkers. Somatic mutations are also biomarkers of effect.

Superimposed on the above are biomarkers that identify individuals who are at increased risk of developing diseases due to environmental exposures. These are biomarkers of susceptibility. Some individuals may have a heightened (or decreased) ability to metabolize xenobiotics to genotoxic intermediates, thereby resulting in increased (or decreased) *in vivo* exposures to equivalent external exposures to environmental genotoxins. Other individuals may have a limited capacity to repair DNA, thereby suffering increased genotoxic effects from equivalent *in vivo* or external exposures to deleterious agents. There are doubtlessly many other factors that determine individual susceptibility, and, therefore, individual risk. Theoretically, biomarkers of susceptibility may identify these individuals.

Costs of Currently Used Biomarker Assays

Since many of the currently available biomarkers have been field tested for human population studies, it is possible to estimate their costs. Table 1 lists the more commonly used biomarkers and approximate unit costs. Several caveats must be noted. First, many or most unit costs include the developmental and validation expenses of early studies. Therefore, actual costs for running the assays *after development and validation* might be as much as one-third to one-half lower. Any large scale study would further reduce expenses. Finally, the exact method used for the assay will determine final cost. For example, biomonitoring for the presence of a single agent or metabolite in body fluids by a well validated HPLC-based method might cost from $35 to $50, whereas the detection of multiple agents on the same sample might cost $150 or more. Similarly, one laboratory estimates that analyses for chromosome aberrations by traditional methods cost approximately $100 per assay, whereas use of newer techniques involving "chromosome painting," while giving better statistical validity, might cost twice as much. Also, although assays for SCE and chromosome aberrations separately are about $100 each, both performed on the same sample would be only about $150.

Given the above caveats, it is still obvious that monitoring biomarkers in humans is not an inexpensive exercise, and that monitoring large populations will be quite costly. The Department of Energy (DOE), which is responsible for the cleanup of the United States nuclear weapons sites, estimates that there are approximately 1,000,000 present and former contractor and subcontractor employees who have worked at DOE sites and who may have been exposed to hazardous environmental agents. A recent law, Public Law 102-484, mandates that at least a portion of these workers should be monitored for exposures to genotoxic agents (National Defense Authorization Act for Fiscal Year 1993, 1992). If one uses this population as an example, the expense of monitoring all of these individuals, using even the most inexpensive assay and reducing the cost by half, would be about $25,000,000. Effective monitoring of such workers will require more than a single assay of an exposure biomarker, and the nature of the hazardous material will often be unknown. Clearly, the goals of such an undertaking must be explicit, the ability to achieve them realistically assessed, and the potential cost savings over "not monitoring" must be carefully weighed in designing any systematic monitoring program for this population.

TABLE 1. ESTIMATED COSTS OF CURRENTLY USED ASSAYS FOR BIOMARKERS

Exposure/Dose Biomarkers	-->	Effect Biomarkers	Cost (unit)*
Chemicals/Metabolites/ Adducts in Body Fluids			$50 (single agent) $150 (multiple agents)
Protein Adducts			$200
	DNA Adducts		$200
	SCE		$100
		Chromosome Aberrations	$100 - $200
		Micronuclei	$100 (lymphocytes)
		Somatic Mutations in Reporter Genes	$100 (GPA) $100 (HPRT)

* Often includes developmental costs; costs would be 1/2 to 1/3 lower for large quantities.

Costs of Medical Evaluation

Workers potentially exposed to hazardous and radioactive substances should receive periodic medical examinations and appropriate testing. For purposes of deriving a first estimate of costs, it is reasonable to include the various cancer screening tests (Table 2).

Screening has been demonstrated to be effective for breast, uterine, cervical, and skin cancers such as melanomas and is probably effective for colorectal, oral and perhaps some kinds of bladder cancer. However, screening has not been shown to be effective for lung cancer or cancers without a localized stage, such as leukemias (National Cancer Institute, 1993). The cost-effectiveness of screening for other cancers (e.g., gastric or hepatic) will depend on specific increased risks due to the particular environmental exposure. Similarly, some special tests (e.g., serum alphaprotein) are not used for the general population in the United States because they are cost-effective only in settings of increased risk due to certain exposures. In addition, there may be some routine tests performed for general medical evaluation, such as complete blood counts. A conservative estimate of the **average** dollar cost of everything is $200 per evaluation, realizing that some tests will be "special" and, therefore, more expensive, but that others will be performed on a less than annual basis.

TABLE 2. COSTS OF MEDICAL EVALUATION AND TESTING PER INDIVIDUAL

Periodic   = Annual
Testing   = Cancer Screening Tests per Examination

        Effective: Breast, Cervix, Melanoma
        Probably Effective: Colorectal, Oral, Bladder
        Possibly Effective: Prostate, Testicular
        Not Effective: Lung, Leukemia

Estimated Annual Costs per Individual:

        $200 for Evaluation and Testing
        $50 for Informing Workers, Counselling, Referrals, Administration

---

In addition to the technical aspects, workers need to be kept informed of the purposes of the evaluations and the results, and appropriate counseling and referral for treatments should be made. These additional requirements will have their own dollar costs which are difficult to determine in a first approximation, but will be estimated as an "overhead" of about $50 per annual evaluation. Other aspects of overhead, such as administration, can be included here. None of these estimates include the costs of treatments.

Although this analysis provides only the crudest estimates of the costs associated with medical evaluations and testing, it can serve to provide a background against which costs of biomonitoring are evaluated. For approximately 1,000,000 workers potentially exposed to hazardous substances, if there is no way to pre-screen this population, the annual cost of medical evaluations will be approximately $250,000,000. Even though the costs of exposure monitoring are high, the **annual** costs of "non-monitoring" such a population are even higher and the total cost of "non-monitoring" will be staggering. Therefore, a scheme can be developed to use biomarkers in the most cost-efficient manner.

USE OF BIOMARKERS

Current Applications

The "biomarker of choice" for human monitoring will depend on the intermediate goal to be achieved. Logically, exposure/dose biomarkers should be the first line assays when this goal is exposure assessment. Effect biomarkers are the assays of choice for deciding if an exposure to an environmental genotoxin has been "processed" and has resulted in genetic damage. Biomarkers of susceptibility are useful in interpreting intra-individual differences in response to environmental hazardous substances and, in some select cases, to estimate individual safety in a particular environment. To use one kind of biomarker to assess a host-environment interaction about which it only indirectly provides information is, at best, to use it inefficiently.

A caveat on the above is that often one kind of biomarker might be used because a more appropriate one is simply not available. Certainly, chromosome aberrations, which must be considered as biomarkers of effect, remain the standard "internal dosimeters" of acute ionizing radiation exposures. However, even in these cases, it would be preferable to have more sensitive biomarkers of exposure, especially for low-dose chronic exposures.

In deciding how to use biomarkers for population monitoring on a large scale, it is useful to distinguish what is currently most reasonable from what might be possible in the future. At present, it would seem that the biomarkers most appropriate for identifying workers exposed to hazardous substances should be the simplest exposure/dose measurements available. These might be assays for agents, metabolites or adducts in body fluids or for specific protein or DNA adducts when the potential offending agents are known, general measures of DNA adducts when the agents are unknown or are complex mixtures and, in either instance, determinations of SCEs. Because chromosome aberrations have been the standard for human population monitoring for several decades, this endpoint should also be determined, if for no other reason than to relate the data developed in this monitoring scenario to the cumulative world-wide experience. If the potential hazardous substance exposures to workers occurred in the remote past decades, measurements of glycophorin-A (GPA) red blood cell variant frequencies may be added to the battery of exposure/dose biomarkers. All of these determinations would cost approximately $600 per monitored worker. Since some choices will be made in different worker pools, the estimate might be lowered to $500 per monitored individual (Table 3).

TABLE 3. PROPOSED BATTERY OF EXPOSURE/DOSE BIOMARKERS

Assays for Chemicals/Metabolites/Adducts in Body Fluids	$50-$150
Protein/DNA Adducts	$200
Chromosome Aberrations/SCE	$150
± GPA Variant Frequency Assays	$50 - $100
	$450 - $600
Estimate ~ $500/Monitored Worker	

Next must come decisions as to whom to study. It is generally agreed that the current state of validation of biomarkers does not justify decisions or assessments regarding individuals, but only regarding groups or cohorts. Therefore, the exposure/dose biomarkers suggested above should be used to study statistically representative samples of cohorts of workers. If group mean results are elevated with respect to appropriate concurrent controls (to be carefully defined prior to study, and itself a controversial issue), the entire cohort represented by the sample must be considered to have been exposed, and must undergo the medical evaluation and testing proposed above. If group mean results are not different from mean results of appropriate controls, the entire cohort represented by the sample will be considered unexposed and will not undergo medical evaluation. However, this aspect of the approach is more difficult to defend. It is this "unexposed" group, of course, that provides the rationale for and savings that result from monitoring.

Again, there will be caveats. If an extremely specific biomarker of exposure/dose is available, it might be possible to make some decisions about the need for medical evaluation from an individual's monitoring results. Also, the question of "outliers" must be addressed. Nonetheless, for gross approximations of probable current costs, and to assess the optimal monitoring scheme for reducing costs of medical evaluations and testing, this general approach provides a reasonable working model.

The costs and savings of this general approach might be illustrated by a vastly oversimplified scenario (Table 4).

TABLE 4. ILLUSTRATIVE SCENARIO

Use of "Exposure/Dose" Biomarkers to Assess Exposures to Hazardous Substances and Reduce Annual Costs of Medical Evaluation and Testing.

1 Million Workers:

    Assume:

        1,000 Cohorts of 1,000 Workers Each
          300 Cohorts Determined to be "Non-Exposed" (Job Description, Ambient Monitoring)
          700 Cohorts "Possibly" Exposed

"Exposure/Dose" Biomarker Assays on Representative Sample of 25 Workers per Cohort

        Cost = $500 × 25 × 700 = $8.75 Million
        Controls & Overhead = $1.25 Million
                        $10 Million

        Assume 20% of Cohorts "Exposed"

Medical Evaluation and Testing Required on 0.2 × 700 × 1,000 = 140,000 Workers
Annual Costs = $250 × 140,000 = $35 Million

For ease of exposition, assume that the 1,000,000 potentially exposed workers exist as 1,000 cohorts of 1,000 workers each. Assume further that 300 of these cohorts can be judged as "non-exposed" because of nature of the job, previous ambient monitoring, etc. Therefore, 700 cohorts of 1,000 workers each will be monitored, using the above suggested $500 package of biomarkers. If 25 individuals from each cohort constitutes a statistically valid sample, the cost of the monitoring would be $8,750,000 (700 × 25 × $500). This total could easily come to $10 million, depending on concurrent controls and administration.

These monitoring costs must be viewed in comparison with potential costs of "non-monitoring." Without monitoring, medical evaluation and testing of the 700,000 workers would cost $175 million annually (which is $75 million less than medical evaluations of the entire 1,000,000 workers, because of decisions made on the basis of job descriptions and ambient monitoring). The savings of monitoring, of course, will depend on the results. Assume that 20% of the cohorts showed elevated mean values for biomarkers of exposure/dose. Then, these entire cohorts, i.e., 140,000 workers, would require the **annual** medical evaluations and cancer testing. This gives an annual cost of $35 million, but provides an annual savings of $140 million over non-monitoring. (Plus, the $75 million that is saved annually will be eliminating 300,000 workers as possibly exposed by job description and ambient monitoring.) Therefore, using these exposure/dose biomarkers in this simplified scenario, monitoring should provide annual savings of $140 million at a one-time cost of $10 million.

Future Applications

It is possible that in the future, biomarkers may be used in different and much more efficient ways. If biomarkers can be validated as surrogates for disease-causing genetic damage, they can be used as indicators of individual (as opposed to population) risk. This would permit medical evaluations and testing to be targeted to individuals at highest risk, and could even guide interventions.

The use of biomarkers discussed thus far has been to determine exposure/dose. However, assessment of individual risk requires identification of genotoxic effects, which is a task for biomarkers of effect. At present, biomarkers of effect encompass chromosome aberrations, micronuclei, and somatic mutations.

Table 5 lists current and potential applications of biomarkers of effect. As noted, they are often used for what should be their secondary purpose, i.e., to determine exposure/dose, with chromosome aberrations being the "standard" for assessing ionizing radiation exposures. Also, the glycophorin-A assay for somatic mutations may be useful for detecting mutagen exposures of the remote past (Kyoizumi et al., 1989). Another somatic mutation assay, that measuring HPRT mutations in T-lymphocytes, may have advantageous features in certain circumstances that make it useful for exposure/dose assessment. Because HPRT mutant cells can be recovered for molecular analyses, determinations of mutational spectra have been proposed to identify specific environmental exposures. The underlying principle is that different genotoxins probably produce different mutational spectra, allowing the nature of the exposure to be inferred from the spectrum produced (Albertini et al., 1990). Thus, this biomarker of effect, which would not be expected to have high sensitivity for exposure/dose detection, may provide specificity in this regard.

TABLE 5. USE OF PRIMARILY "EFFECT" BIOMARKERS IN HUMANS

1) FOR ASSESSING "EXPOSURE/DOSE"

    o Chromosome Aberrations for Ionizing Radiation
    o GPA Variant Frequencies for Remote Exposures
    o HPRT Mutant Frequencies and Mutational Spectra in Placental Blood for Community Exposures*
    o HPRT Mutational Spectra for "Specificity"*

2) FOR ASSESSING "EFFECTS"

    o Determine Genotoxic Effects of Low Dose Exposures (Known), e.g., HPRT Mutations and Butadiene
    o Monitor Chemopreventive Interventions - Chronic or Recurrent Exposures
    o Define Individual Risk for Developing Genotoxic Disease*

    *To be demonstrated (Research Applications)

HPRT T-cell mutations do provide a sensitive indicator of exposure/dose in the fetus. Mutant frequencies at this locus are very low during fetal life, and the spontaneous mutational spectrum is characteristic (McGinniss et al., 1989, 1990; Fuscoe et al., 1991). Therefore, monitoring community exposure by measuring fetal mutations, as determined in newborn placental cord blood samples, may be feasible. These applications remain to be fully validated by research.

Assessment of genotoxic consequences, however, is the primary purpose of biomarkers of effect, and it is in this role that they might allow prediction of individual risk. Table 5 lists three such uses for effect biomarkers. One is to determine if low dose exposures to known genotoxins do, in fact, have genotoxic consequences. This information could help to determine if "safe" mutagen exposures are truly without genotoxic effects or, at least, if these effects are less harmful than those resulting from tobacco use. Two recent but controversial studies have suggested that exposures to exogenous mutagens, i.e., radon and butadiene, at levels thought to be safe, might have had genotoxic consequences as measured by HPRT mutations (Ward et al., 1994; Bridges et al., 1991).

Another use for biomarkers of effect might be to monitor chemoprevention trials using chemo- or radio-protective agents in situations where individuals are undergoing known genotoxic exposures. This is primarily a medical application, but it could be useful in the future in cases of accidental mutagen exposures.

Finally, the use for biomarkers of effect most relevant to cleanup workers may be as indicators of individual risk. If they can be used in this way, they will have enormous value for preventive medicine. This use, however, requires their validation as disease surrogates, an area which requires further research.

At present, there is suggestive evidence that biomarkers of effect do predict individual risks for developing genotoxic diseases (Table 6). Early reported results of a prospective follow-up Nordic study indicate that individuals with the highest frequencies of structural chromosome aberrations are at an elevated cancer risk (Sorsa et al., 1992). Therefore, at least this effect biomarker may predict individual health risk. Additionally, two lines of evidence relate indirectly to the question of individual risks predicted by effect biomarkers. The first, deriving from animal studies, is the concomitant decrease in neutron irradiation-associated premature deaths, induced cancers and *in vivo* lymphocyte HPRT mutations in mice (Grdina et al., 1991; 1992). This, however, was a population level association so individual risks could only be inferred. On the mechanistic level in humans, there is now a good deal of evidence indicating that carcinogenic mutagenic mechanisms are "captured" in *in vivo* reporter gene mutations. These include somatic recombinations, which are detected by glycophorin-A and HLA mutations (Langlois et al., 1989; Turner et al., 1988), and illegitimate V(D)J recombinase-mediated mutations, which are detected by the HPRT system (Fuscoe et al., 1991; 1992). All of these are known to be important when they occur in cancer genes. Although encouraging, this last line of evidence is currently theoretical.

Overall, therefore, there remains a need to validate biomarkers of effect in terms of their ability to predict, in the individual, the probable occurrence of subsequent genotoxic disease.

TABLE 6. SUGGESTIVE EVIDENCE THAT "EFFECT" BIOMARKERS MAY PREDICT
INDIVIDUAL RISKS FOR DEVELOPING GENOTOXIC DISEASE

---

o   Chromosome Aberrations and Cancers - Nordic Study

o   Radiation Induced Cancers and HPRT Mutations in Mice Both Reduced by
      Radioprotective Agent

o   Carcinogenic Mutagenic Events "Captured" in Reporter Genes
      Somatic Recombination: GPA and HLA
      V(D)J Recombinase Events: HPRT

Validation of Effect Biomarkers for Individual Risk

True validation of effect biomarkers as predictors of individual risk
will require collaboration among genetic toxicologists, epidemiologists,
and physicians. Clearly this will require a good deal of thought and
organization, but costs should be moderate as such studies go, and
the results, if validation can be achieved, will be well worth the effort.
Perhaps the most important unresolved question concerning the use of
biomarkers of effect for human biomonitoring is their relevance to disease
outcomes.

Table 7 lists three approaches for validating biomarkers of effect
in terms of their disease prediction potential. The first is to add *in
vivo* mutagenicity studies to ongoing trials of human cancer chemoprevention. If the chemopreventive agent being tested acts by preventing
DNA damage, it should concomitantly reduce *in vivo* reporter gene
mutations as it decreases cancer occurrences. When such trials are conducted under conditions of continuing mutagen/carcinogen exposures,
as in cancer patients receiving radio- or chemotherapies, a strong correlation between reduction of secondary cancers and decreases of induced
in vivo mutant frequencies would indicate that the latter predict the
former.

Indirect evidence that biomarkers of effect predict cancer outcomes
could also come from studies that take advantage of the technology for
directly detecting cancer mutations *in vivo* in humans. A strong intra-

individual correlation between mutations in reporter genes and cancer genes in individual cancer patients under treatment will be strong evidence that the former mutations are valid surrogates for the latter, and suggest that the reporter gene events are useful predictors of subsequent cancers.

TABLE 7. APPROACHES FOR "VALIDATING" HUMAN "EFFECT" BIOMARKERS AS PREDICTORS OF "INDIVIDUAL" RISKS FOR DEVELOPING GENOTOXIC DISEASE

o   Covariation of Biomarker Responses and Individual Outcomes in Chemoprevention Intervention Trials Following Mutagen/Carcinogen Exposures

o   Correlation of Biomarker Responses (e.g., Somatic Mutation in Reporter Gene) with Mutations in Cancer Genes

o   Retrospective Case-Control Studies Relating Biomarker Responses to Disease Outcomes, Using Repositories and Cryopreserved Materials from Exposed Individuals

The most relevant studies for validating biomarkers of effect as predictors of subsequent disease outcomes in individuals would be those that directly test this association. This may be accomplished more easily than expected, especially with the use of a repository for cryopreserved tissue samples, primarily blood, from individuals exposed to known, large doses of mutagens/carcinogens (Albertini, 1982). Most exposed individuals would be cancer patients receiving chemotherapy or radiotherapy treatments, but samples from survivors of environmental disasters could also be used if exposures are known. Ideally, multiple samples would be obtained from the same individuals who would each be kept under medical surveillance. (Ongoing medical surveillance is usual in such instances.) Outcomes in terms of disease occurrence would be regularly entered into the repository database so that, in time, there would be a longitudinal medical history for each individual who donated one or more post-exposure blood sample. (Pre-exposure samples would probably also be available for the cancer patients.) No testing of samples for biomarkers need be performed until there has been sufficient occurrence of disease to permit a statistically valid assessment. For purposes of initial validation, disease outcomes will be limited to cancer and, within this disease category, to leukemia.

Leukemia is a cancer that may occur within a few years of carcinogen exposure. At some point in the follow-up process, a panel of epidemiologists, physicians, geneticists and statisticians would design nested "case-control" studies to validate the biomarkers studied as surrogates for disease outcomes by relating leukemia outcomes among all exposed individuals to the results of biomarker analyses. The relative risks ("odd ratios") for developing leukemia among exposed individuals who share different levels of mutant frequency elevations would be determined relative to exposed individuals who do not show such elevations. Of course, exposed individuals without elevations of mutant frequency values must be shown also *not to have* increased leukemia occurrences relative to unexposed individuals. All testing would, of course, be blind. Obviously, this exercise would not be done to confirm the mutagenicity/carcinogenicity of agents to which the individuals were exposed, or even to determine doses received. Rather, it would be done to determine if mutagenicity tests have validity as predictors of leukemia or other cancer outcomes in individuals. Over a five- to ten-year period, this research should provide answers to the question of validity of these biomarkers as individual risk predictors. The answer will be statistical, i.e., relative risks or "odds-ratios." Only after the biomarkers are so validated will we be able to use them to guide the medical evaluation and testing required for individuals exposed to hazardous substances.

Value of Biomarkers Determining Individual Risk Assessments

A proposed scenario for using effect biomarkers to guide the medical evaluation and testing of workers exposed to hazardous substances is provided in Table 8. It is emphasized that this scenario is hypothetical and oversimplified, and serves only to illustrate how biomarkers of effect might make medical evaluations and testing more effective and save costs.

The biomarkers of effect (e.g., somatic reporter gene mutations) used for determining individual risks must first be validated, using methods similar to those outlined above. As noted, results of validation studies would be in terms of relative risks (RR) or, more precisely, odds ratios. Assume, for example, that validation studies demonstrated an RR of 2.0 for developing cancer in exposed individuals with adjusted mutant frequency values (for some reporter gene mutation) 3 times that expected. Based on this outcome, a decision would then have to be made regarding the medical surveillance of workers exposed to the same hazardous substance(s). Assume further that medical evaluation and testing is not warranted for individuals with RRs less than 2.0. This may be justified by other than cost considerations. For example, pre-test likelihoods of developing cancer may be far too small in some instances for the medical evaluation and testing to be useful.

TABLE 8. USE OF "EFFECT" BIOMARKERS TO REDUCE ANNUAL COSTS
OF MEDICAL EVALUATIONS AND TESTING FOR WORKERS EXPOSED
TO HAZARDOUS SUBSTANCES

Decision:	No Medical Evaluation and Testing if RR for Developing Cancer (Leukemia) <2.0
Results of "Validation" Studies:	R.R. for Developing Cancer >2.0 for Adjusted Mutant Frequency Values > 3 × Expected
Future Scenario:	140,000 Workers Assessed as "Exposed" 10% Adjusted Mf Values > 3 × Expected
Therefore:	14,000 Workers Require Annual Medical Evaluation and Testing = 14,000 × $250 = $3.5 Million Annual Cost
	Cost of Monitoring $28 Million
	$31.5 Total Cost but Subsequent Annual Costs Reduced by $31.5 Million

The future scenario for efficient risk management of the 1,000,000 workers potentially exposed to hazardous substances we have taken as a model for this analysis might then be as follows. As shown in Table 4, after considerations of job description and ambient monitoring, and after biomonitoring representative samples of workers in the remaining "potentially" exposed cohorts using exposure/dose biomarkers, 140,000 workers were assessed as exposed. The annual cost of medical evaluation and testing of this group would be $35 million. However, studies using effect biomarkers (two markers, or one in duplicate) (e.g., somatic mutations) at a cost of $200 per individual (Table 1) might reveal only 10% of this group, or 14,000 workers, to have adjusted mutant frequency values 3 times that expected on at least one study. These are the workers with RRs of 2 times background for developing cancer. This biomonitoring study using somatic mutations costs $28 million, but it reduced the size of the population requiring medical evaluation and testing from 140,000 to 14,000. The cost of the medical studies in this group, at $250 per individual (Table 2), has been reduced from $35 million (Table 4) to $3.5 million (14,000 × $250). The total first year cost of the biomonitoring plus medical evaluations and testing was $31.5 million, for a savings of only $3.5 million. However, assuming that the exposure to the hazardous substance is now over, biomonitoring need not be repeated. Therefore, the annual cost for medical evaluation and testing will remain at $3.5 million, for an annual savings of approximately $31.5 million.

Can effect biomarkers really be used in this way? This depends on the results of validation studies. Certainly, they *CANNOT* be used in this way at present without the requisite validations. There should be no misunderstanding on this point. However, the enormous cost savings to be realized if biomarkers of effect can be used in this way and the more effective medical evaluation and testing that will be possible because of knowledge of individual risks would make investment in the research necessary to validate them well worth the cost.

CONCLUSIONS

The costs of biomonitoring a large population of nuclear weapons sites cleanup workers who are potentially exposed to hazardous substances will be high. However, the annual costs of "non-monitoring" and relying only on medical evaluation and testing would probably be even greater. Biomarkers may be useful in reducing these costs. A hypothetical scenario in which decisions based on job descriptions and ambient monitoring shows how the potentially exposed population of workers can be reduced substantially. Biomonitoring of this residual population using exposure/dose biomarkers is currently possible. A scheme can be developed in which representative samples of cohorts of potentially exposed workers are tested with a battery of exposure/dose biomarkers. This approach could further reduce the numbers of workers who must undergo annual medical evaluation and testing. It is possible that, in the future, further biomonitoring of workers assumed to have been exposed using biomarkers of effect will allow assessments of individual risks. This has the potential for further reducing the numbers of workers who will require annual medical evaluation and testing and for making the medical surveillance that is conducted more effective. This last mentioned use for effect biomarkers will depend on the result of validation studies, which should be a high priority research investment.

(Supported by DOE grant # DE-FG02-87ER60502 and NCI grant # CA30688. DOE and NCI support does not constitute an endorsement of the views expressed herein.)

REFERENCES

Albertini, R. J. 1982. Studies with T-lymphocytes: An approach to human mutagenicity monitoring. Pp. 393-412 in: Bridges B. A., B. E. Butterworth and I. B. Weinstein, eds. Banbury Report: Indicators of Genetic Exposure, Vol. 13, Cold Spring Harbor Laboratory, NY.

Albertini, R. J., J. A. Nicklas, J. P. O'Neill, and S. H. Robison. 1990. *In vivo* somatic mutations in humans: Measurement and analysis. Annual Reviews of Genetics 24:305-326.

Bridges, B., J. Cole, C. F. Arlett, M. H. L. Green, A. P. W. Waugh, D. Beare, D. L. Henshaw, and R. D. Last. 1991. Possible association between mutant frequency in peripheral lymphocytes and domestic radon concentrations. Lancet 337:1187-1189.

Committee on Biological Markers (National Research Council). 1987. Biological markers in environmental health research. Environmental Health Perspectives 74:3-9.

Fuscoe, J. C., L. J. Zimmerman, M. L. Lippert, J. A. Nicklas, J. P. O'Neill, and R. J. Albertini. 1991. V(D)J recombinase-like activity mediates HPRT gene deletion in human fetal T-lymphocytes. Cancer Research 51:6001-6005.

Fuscoe, J. C., L. J. Zimmerman, K. Harrington-Brock, L. Burnette, M. M. Moore, J. A. Nicklas, J. P. O'Neill, and R. J. Albertini. 1992. V(D)J recombi-nase -mediated deletion of the HPRT gene in T-lymphocytes from adult humans. Mutation Research 283:12-20.

Grdina, D. J., B. A. Carnes, D. Grahn, and C. P. Sigdestad. 1991. Protection against late effects of radiation by S-2-(3-aminopropylamino)-ethyl-phosphoro-thioic acid. Cancer Research 51:4125-4130.

Grdina, D. J., Y. Kataoka, I. Basic, and J. Perrin. 1992. The radioprotector WR2721 reduces neutron-induced mutations at the hypoxanthine-guanine phosphoribosyltransferase locus in mouse splenocytes when administered prior to or following irradiation. Carcinogenesis 13:811-814.

Kyoizumi, S., N. Nakamura, M. Hakoda, A. A. Awa, M. A. Bean, et al. 1989. Detection of somatic mutations at the glycophorin-A locus in erythrocytes of atomic bomb survivors using a single beam flow sorter. Cancer Research 49:581-588.

Langlois, R. G., W. L. Bigbee, R. H. Jensen, and J. German. 1989. Evidence for increased in vivo mutations and somatic recombination in Bloom's syndrome. Proceedings of the National Academy of Sciences (USA) 86:760-674.

McGinniss, M. J., J. A. Nicklas, and R. J. Albertini. 1989. Molecular analyses of in vivo HPRT mutations in human T-lymphocytes. IV. Studies in newborns. Environmental Molecular Mutagenesis 14:229-237.

McGinniss, M. J., M. T. Falta, L. M. Sullivan, and R. J. Albertini. 1990. In vivo HPRT mutant frequencies in T-cells of normal human newborns. Mutation Research 152:107-112.

National Cancer Institute. 1993. PDQ (Physician Data Query) Database. Cancer Screening Summaries, 9/14/93.

National Defense Authorization Act for Fiscal Year 1993. 1992. (Public Law 102-484). Section 3162. Program to Monitor Department of Energy Workers Exposed to Hazardous and Radioactive Substances. 102d Congress. October 23, 1992. Washington, D.C.

Sorsa, M., J. Wilbourn, and H. Vainio. 1992. Human cytogenetic damage as a predictor of cancer risk. Pp. 543-554 in: Mechanisms of Carcinogenesis in Risk Identification. H. Vainio, P.N. Magee, D.B. McGregor and A.J. McMichael, eds., International Agency for Research on Cancer, IARC, Lyon, France.

Turner, D. R., S. A. Grist, M. Janatipour, and A. A. Morley. 1988. Mutations in human lymphocytes commonly involve gene duplication and resemble those seen in cancer cells. Proceedings of the National Academy of Sciences (USA) 85:3189-3192.

Ward, J. B., M. M. Ammenheuser, W. E. Bechtold, E. B. Whorton Jr., and M. S. Legator. 1994. HPRT mutant lymphocyte frequencies in workers at a 1,3-butadiene production plant. Environmental Health Perspectives (in press).

Biomarkers and Occupational Health: Progress and Perspectives. 1995. Pp. 89-102

Biological Monitoring of Exposure to Industrial Chemicals

Vera Fiserova-Bergerova (Thomas)

In the past, inhalation exposure was considered the main threat to workers' health, and exposure to toxic chemicals was evaluated by measurements of airborne contaminants in the workplace. Such measurements provide information on the source of the air pollutant(s) (area samples) or on the inhalation exposure of individual workers (personal samples). The risks from other sources of exposure were recognized for only a few chemicals, mainly those that induce toxicity upon contact with skin. This practice endures until today.

Air monitoring, however, fails to provide information on the total exposure of the worker, which is a conglomerate of exposures via multiple absorption routes (inhalation, dermal, gastrointestinal) and from multiple sources (occupational, nonoccupational, environmental, dietary). As the reference values for permissible inhalation exposures are being reduced to provide protection against preclinical changes, the contribution of inhalation exposure to the total exposure diminishes and the contribution from other routes and sources gains importance. At present, biological monitoring is the only available tool for the evaluation of total exposure.

The measures of total exposure are the levels of the chemical or its metabolite(s), or biochemical change(s) (further referred to as determinants) induced by the exposure, in biological specimens (urine, blood, exhaled air, and appendages) collected from exposed workers. Biological effects usually correlate better with biological levels than with airborne concentrations. In order to promote biological monitoring, in 1982 the American Conference of Governmental Industrial Hygienists (ACGIH) established a committee on Biological Exposure Indices (BEI) and charged it with recommending reference values for biological levels of determinants. A similar committee was established in Germany for Biologische Arbeitsstofftoleranzwerte (BAT).

Biological levels of determinants depend on the following variables:

- Exposure: degree and duration, workload and co-exposure.
- Sampling: timing and duration of sampling (depend on the kinetic profile of the compound).
- Background levels: endogenic, environmental, and dietary factors.
- Physiological parameters of workers: body fat, clearance, availability of metabolizing enzymes, and binding sites.

To provide reference values for biological monitoring, these variables must be defined. BEI values are determined as biological levels of the determinant in biological specimens collected from a healthy reference man (170 cm, 70 kg, about 12% body fat) occupationally exposed (8 hours per day, 5 days per week) to a chemical or a specified mixture while performing work with a moderate energy expenditure (50 W, pulmonary ventilation about 20 L/min). The sampling time for each BEI is specified according to the elimination half-life (or half-lives) of the determinant.

The main considerations in establishing BEIs are:

- Availability of a reliable method (considerations include sensitivity, specificity, reproducibility, form of determinant, stability of sample).
- Sampling time and sampling techniques (critical for a chemical with a short half-life or multiphasic elimination).
- Accumulation (critical for a chemical with a long half-life or multiphasic elimination).
- Background levels (basing BEI on the increment caused by occupational exposure eliminates the variability induced by environment and diet).
- Specificity of the indicator.
- Correlation of the level of the indicator with the degree of exposure.
- Information on exposure (determinants with a short half-life provide information on recent exposures; determinants with long half-lives provide information on past exposures).
- Information on biological effect (is it related to level or dose?).
- Variability in physiological factors affecting the biological levels

(body build and workload, inhibition or induction of metabolism by other xenobiotics or disease, and ethnic or genetic factors).

- Effect of co-exposure to other xenobiotics (exposure to a mixture of industrial chemicals, consumption of alcoholic beverages, smoking, medication).
- Available data base (volume and reliability of data).

The main components of the BEI data base are:

- Animal studies and case reports (to establish the effect of the chemical and mechanism of the effect).
- Laboratory studies with volunteers (to generate pharmacokinetic data).
- Field studies linking biological levels to health effect.
- Field studies comparing results of air monitoring in the workplace and levels of determinants in specimens collected from workers.
- Epidemiological/environmental studies (to establish background levels of determinants in occupationally unexposed population).
- Pharmacokinetic models (to adjust data from laboratory studies to exposure in the workplace considering workload and timing; to extrapolate a BEI for occupational exposure to the Threshold Limit Value-Time Weighted Average (TLV-TWA); and to lay grounds for establishing sampling time).

Interaction Between BEI and TLV Committees

At present, BEIs have been established (or are on the list of intent to be established) for 34 industrial chemicals (22 organic solvents, 6 metals, 6 pesticides and others). They represent the levels of determinants that are most likely to be observed in specimens collected from a healthy worker who has been exposed to the chemical to the same degree as a worker exposed at TLV-TWA (by inhalation only). The exceptions are the BEIs for chemicals for which TLVs are based on protection against nonsystemic effect(s) (e.g., irritation or respiratory impairment). The BEIs, extrapolated from the TLV on a pharmacokinetic basis, are compared with the results in field studies linking biological levels with biological effect. Whenever the extrapolated BEI exceeds the no-effect level, the proposed BEI and TLV are revised by both the BEI and TLV Committees. Whenever a change of the TLV is proposed, revision of

the BEI is initiated (currently lead, benzene, and toluene). Current
BEIs are published each year (usually in September) by ACGIH in the
popular TLV-BEI booklet. The first BEIs appeared in the 1984-1985
edition. ACGIH also publishes the bases for the BEIs in the same year
as they are recommended.

Criteria for Safety

As the studies of effects of chemicals on humans are refined, lower
no-effect levels are recognized. Evaluation of the significance of preclinical changes, however, becomes questionable (e.g., occurrence of low
levels of some proteins and enzymes in urine after exposure to heavy
metals; changes in cognitive and performance tests). Since there are
no objective criteria for how to evaluate these "questionables," the decisions are made by a vote of the experts on the BEI Committee (10
members) and TLV Committee (20-25 members) and by the member-
ship of ACGIH. Safety factors are also a matter of debate and vote,
consideration being given to the gravity and reversibility of the effect.

Implementation of BEIs

Industrial hygienists should use biological monitoring to substan-
tiate air monitoring, to test the efficacy of personal protective equipment, to determine the potential for absorption via the skin and the
gastrointestinal system, and to detect nonoccupational exposures. BEI
values and documentations provide guidelines for exposure evaluation.
BEIs do not indicate a sharp distinction between permissible (harmless)
and unacceptable (hazardous) exposures. The indicated sampling time
must be observed if a BEI is applied. Multiple sampling is necessary to
reduce the effects of variable factors. The average value should be compared with the BEI, and the distribution of measured values should be
observed. Measurements significantly exceeding the BEI should be repeated immediately and, if confirmed, the cause should be investigated.
Monitoring should be done on a regular basis, and measurements should
be evaluated historically with respect to the health of the worker(s) as
well as with respect to workplace safety. Special consideration should
be given to co-exposure to other industrial chemicals and xenobiotics.
Regional and ethnic differences in diet, environment, lifestyle, and enzyme activity should be observed. Specific issues have been addressed
by the members of the BEI Committee in a series of 8 articles on "Biological Monitoring" which were approved as the Committee policy and

published during 1989 to 1993 in Applied Industrial Hygiene (renamed Applied Occupational and Environmental Hygiene).

CONSIDERATIONS IN BIOLOGICAL MONITORING DATA EVALUATION

Implementation of BEIs to Exposures to Mixtures of Chemicals

Workers are usually exposed in the workplace to more than one chemical. These chemicals can interact in the body by competing for binding sites and enzymes (pharmacokinetic effects) or by interacting in the development of toxicity or altering some physiological function (pharmacodynamic effects). The TLVs for compounds used in mixtures are adjusted to protect against toxicodynamic effects. Similar rules are accepted for adjustments of BEIs. For the independent toxicodynamic effects, no adjustment is required. For the additive toxicodynamic effect, the following biological hazard index "k_b" should be implemented:

$$k_b = \frac{MC_1 - BC_1}{BEI_1 - BC_1} + \frac{MC_2 - BC_2}{BEI_2 - BC_2} + \dots \frac{MC_n - BC_n}{BEI_n - BC_n} \quad (Equation\ 1)$$

where MC and BC are, respectively, the measured and background concentrations of the exposure indicators. The subscripts denote the exposure indicators for individual chemicals. By analogy with the hazard index for air monitoring, the biological hazard index should be smaller than 1, or the value of each ratio in Equation 1 should be approximately the same as the value of the ratio of measured concentration in the air to the TLV-TWA. Further reduction may be necessary if the interaction has a significant toxicokinetic effect on the biological levels of the determinant. For potentiation and synergistic effects, there are no general rules on how to adjust the reference values for biological monitoring. The evaluation of exposure safety, which must be determined individually, is left to the discretion of professionals. In the case of antagonistic effects, no adjustment is required.

The selection of determinants for monitoring of exposure to a mixture requires the following special considerations:

- Specificity of the exposure indicator (the determinant should be the product of only one component of the mixture).

- Biological levels of the indicator should ideally be unaffected by the co-exposure.

- In the absence of specific indicators unaffected by co-exposure, potentially toxic products of exposure should be measured. The measurements provide information on the overall health risk to the worker, but are not necessarily the quantitative measure of exposure.

Sample Collection

Proper sampling and storage of biological specimens are critical for quantitative evaluation of exposure. Special attention should be given to:

- Avoid contamination (sampling containers, air in sampling place, sampling procedures).

- Avoid sample deterioration during storage and transportation (decomposition of biological matrix, evaporation of volatile determinants, adsorption to the walls of containers, and decomposition of the measured determinant).

- Accurate recording of sampling time (timing with respect to the start, peak, and end of exposure; sampling duration).

Since the measurements represent TWA concentrations in biological specimens during the sampling period, the sampling time must be considered in data evaluation.

The causes of the sampling time effects are:

- Inconsistency of *actual biological levels* (their rising and declining depend on the toxicokinetic profile of the determinant and duration of exposure and post-exposure periods).

- Biases introduced by *duration of sampling period* (critical for samples collected during rapid concentration changes).

Actual biological levels increase during exposure and start to decline, after the end of exposure. The rate of this decline is defined by the half-life of the determinant. The main concerns in the interpretation of biological levels are demonstrated by using determinants with very long and very short half-lives. When the half-life of a determinant is in the magnitude of weeks such as in the case of heavy metals:

- The main cause of concern is accumulation.

- The sampling time in relation to the start (or end) of exposure is not critical, since the changes in biological level are very slow. However, samples collected after a few months of exposure are the most valuable for evaluation of workplace safety.

- The measurements that reflect the dose retained during long-lasting exposure usually relate to chronic health effects.

When the half-life of a determinant is in the magnitude of minutes or a few hours such as in the case of volatile organic solvents and some of their metabolites:

- The main cause for concern is rapid changes in biological levels. Sampling duration is critical, especially at the beginning and end of exposure.

- Sampling time in relation to the start (or end) of exposure is very critical and is defined for each BEI.

- The measurements that reflect the exposure immediately before sampling (half-life in minutes) or during the shift (half-life in hours) usually relate to acute health effects.

Sampling duration

Samples can be taken as a bolus (blood, single breath) or over a period of time (urine, multiple breath samples, hair, nails). Since the measured concentration reflects the TWA concentration in the body during the sampling period, the measurements depend on the rising or declining of the concentration during the sampling period, i.e., on the half-life. Tables 1 and 2 demonstrate the effect of the sampling period on measured concentrations for determinants with half-lives of 5 minutes, 5 hours, and 1 day. Figure 1 demonstrates why sampling time is best defined by the middle of the sampling period.

A Case of Sampling Bias: Exhaled air monitoring of volatile compounds.

Pulmonary uptake and wash-out are multiphasic processes with three distinguishable half-lives: one, in the magnitude of minutes, which dominates shortly after the start or end of exposure; a second, in the magnitude of hours, which dominates up to the second day after the start or end of exposure; and a third, with a half-life in the magnitude of days, which dominates later on, when the steady state is approached. Therefore, monitoring of exhaled air depends on the choice of sampling time and technique. Below are examples of the choices.

TABLE 1. EFFECT OF SAMPLING DURATION ON POST-EXPOSURE BIOLOGICAL
LEVELS OF DETERMINANTS WITH DIFFERENT HALF-LIVES

$t_{1/2}$ = 5 min.		$t_{1/2}$ = 5 hrs.		$t_{1/2}$ = 1 day	
Sampling period	Measured conc.	Sampling period	Measured conc.	Sampling period	Measured conc.
bolus	100	bolus	100	bolus	100
5 min.	72	1 hr.	93	1 hr.	98.6
10 min.	54	2 hrs.	87	2 hrs.	97.2
15 min.	42	3 hrs.	82	4 hrs.	94.4
20 min.	34	4 hrs.	77	6 hrs.	91.8
25 min.	28	5 hrs.	72	8 hrs.	89.3
30 min.	24	10 hrs.	54	12 hrs.	84.5

Concentration measured in the instantaneously collected sample (bolus) equals 100. TWA concentrations are given for samples collected over the indicated period. (The duration of sampling profoundly affects the measured concentrations of determinants with short half-lives (e.g., 5 min.) and has little effect on measurements of determinants with long half-lives [e.g., 1 day]).

Choice of sample matrix:

• Total exhaled air (includes air from dead space).

• End exhaled air (mainly alveolar air, which accounts for about two thirds of total exhaled air).

• End exhaled air after apnea (alveolar air).

• Forced exhaled air.

• Mode of respiration (combination of inspiration by nose or mouth and expiration by nose or mouth).

TABLE 2. EFFECT OF SAMPLING TIME ON BIOLOGICAL LEVELS OF
DETERMINANTS IN SPECIMENS COLLECTED DURING EXPOSURE.

Sampling period min.	$t_{1/2} = 5$ min. Concentration actual	mean	$t_{1/2} = 5$ hrs. Concentration actual	mean
5	50	28	1.1	0.6
10	75	46	2.3	1.1
30	98	76	6.7	3.4
45	100	84	9.9	5.0
60	100	88	12.9	6.6
120	100	94	24.2	12.7
240	100	97	42.6	23.2
300	100	98	50.0	27.9
390	100	98	59.4	34.0
480	100	98	67.0	39.6

Concentrations given for specimens collected during exposure either as bolus at the end of
sampling periods (actual) or during the indicated sampling periods (mean), assuming steady
state concentration of 100. (Measurements at the beginning of exposure are more affected by
sampling time than those at the end of exposure.)

The choice of sample matrix affects the outcome of measurements because of two variables, namely, the dead space and the absorption and
reactivity in respiratory airways.

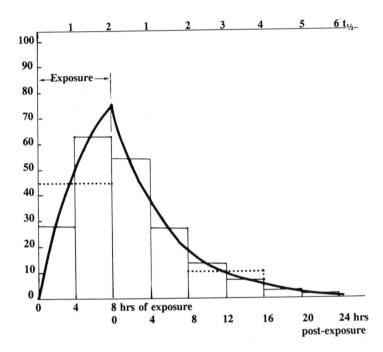

FIGURE 1. Effect of sampling time on measurements of determinants in samples collected during exposure and post-exposure periods. A determinant with a half-life of 4 hours was measured in samples collected during and following an 8-hour exposure. Samples were collected frequently as bolus (e.g., blood) or over 4-hour intervals (e.g., urine). The concentration to be reached at steady state equals 100 (coordinate). Time is given in hours from the beginning or end of exposure (lower abscissa) or in the number of half-lives (upper abscissa). Concentration changes are most rapid immediately after the beginning of exposure and decline with time.

Choice of sampling period:

- Single breath (in test tube).
- Multiple breaths (in large container or in absorption tubes).

Measurements in breath samples collected when the phase with a short half-life dominates are strongly affected by sampling time (Figure 2). Biases introduced by sampling duration are discussed above.

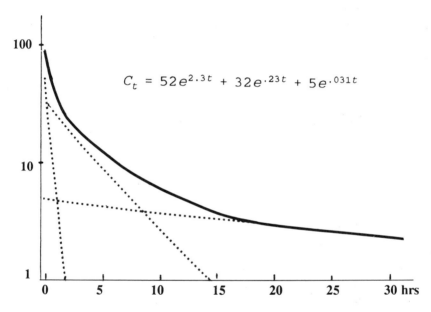

$$C_t = 52e^{2.3t} + 32e^{.23t} + 5e^{.031t}$$

FIGURE 2. Triphasic pulmonary wash-out of volatile compounds. Wash-out is triphasic, with half-lives of 20 minutes, 3 hours, and 22 hours, as a function of alveolar ventilation, perfusion rates, tissue volumes, and biosolubility. The wash-out (bold line) is the sum of the three exponentials (dotted lines). In the semilogarithmic plot, abscissa shows hours after the end of exposure; coordinate shows percentage of original concentration at the end of exposure. (Sampling time is critical during the first post-exposure hour, when processes with a half-life of 20 minutes dominate. In samples collected after 17 hours, slow processes dominate and measurements are affected very little by sampling time.)

Choice of sampling technique:

- **During exposure, inhaling the contaminated air.** If the determinant is the inhaled compound, the measurements are deeply affected by the current concentration in the ambient air. Biological monitoring is usually unnecessary, because it provides no information that cannot be obtained by ambient air monitoring.

- **During exposure, inhaling purified air.** The measurements in samples, usually collected by a mask equipped with a filter, reflect the degree of exposure immediately before the filter application. They are significantly affected by the duration of sampling (see Table 1). Monitoring can, however, be very useful to test the effectiveness of respirators.

- **Samples collected shortly after exposure while inhaling clean air.** Measurements which are an indicator of concentrations at the end of exposure are profoundly affected by sampling time (see Table 1 and Figure 2).

- **Samples collected 10 hours or later after the end of exposure (preshift samples).** The measurements correlate with average exposures on the previous days. Timing is not critical because the sampling takes place during the slow elimination phase. However, since the concentrations reflect the washing-out from depot in the deep compartment (fat), the measurements are affected by post-exposure activity and body build of the worker. Background levels and environmental contamination can interfere.

CONCLUSIONS

Exhaled air monitoring appears to be an ideal noninvasive method for the monitoring of recent as well as past exposures to chemicals. However, a review of the literature reveals that the inconsistency of sampling techniques introduces biases in data interpretation. Therefore, the BEI Committee recently reviewed its policy on exhaled air monitoring and withdrew the BEI for a number of determinants until a uniform methodology for exhaled air sampling is accepted. In the meantime, exhaled air monitoring is, with a few exceptions, recommended as a semi-quantitative screening test.

RECOMMENDED LITERATURE

American Conference of Governmental Industrial Hygienists, Inc. 1993. Threshold limit values for chemical substances and physical agents and biological exposure indices 1993-1994. ACGIH, Cincinnati, OH.

American Conference of Governmental Industrial Hygienists 1991. Documentation of TLVs and BEIs. ACGIH, Cincinnati, OH.

Biological Monitoring of Exposure to Industrial Chemicals. 1990. Proceedings of the United States-Japan cooperative seminar on biological monitoring. V. Fiserova-Bergerova and M. Ogata, eds. American Conference of Governmental Industrial Hygienists Inc. Cincinnati, Ohio.

Biological Tolerance Values for Working Materials. VII. Significance and usage of BAT values. 1993. Pp 123-132 in: Maximum Concentrations at the Workplace and Biological Tolerance Values for Working Materials 1993. Report No. 29, Commission for the investigation of health hazards of chemical compounds in the work area, VCH, Weinheim, Germany.

Droz, P.O. 1985. The use of simulation models for setting BEIs for organic solvents. Annals of the American Conferences of Governmental Industrial Hygienists 12:339-350.

Droz, P.O. 1989. Biological monitoring I: sources of variability in human response to chemical exposure. Applied Industrial Hygiene 4: F20-F24.

Droz, P.O., M. Berode, and M. M Wu. 1991. Evaluation of concomitant biological and air monitoring results. Applied Industrial Hygiene 6:465-474.

Droz, P.O., and V. Fiserova-Bergerova. 1992. Biological Monitoring VI: Pharmacokinetic models used in setting biological exposure indices. Applied Occupational and Environmental Hygiene 7:574-680.

Droz, P.O., M. M. Wu, W. G. Cumberland, M. Berode. 1989. Variability in biological monitoring of solvent exposure. I. Development of a population physiological model. British Journal of Industrial Medicine 46:447-460.

Droz, P.O., and M. M. Wu. 1990. Biological Sampling Strategies. Pp. 251-270 in: Exposure Assessment for Epidemiology and Hazard Control. S. M. Rappaport and T. J. Smith, eds. Lewis Publishers, Chelsea, MI.

Fiserova-Bergerova, V. 1987. Development of biological exposure indices (BEIs) and their implementation. Applied Industrial Hygiene 2:87-92.

Fiserova-Bergerova, V. 1987. Simulation model as a tool for adjustment of biological exposure indices to exposure conditions. Pp. 29-57 in: Biological Monitoring of Exposure to Chemicals: Organic Compounds. M. H. Ho and H. K. Dillon, eds., John Wiley & Sons, Inc., New York.

Fiserova-Bergerova, V. 1990. Application of toxicokinetic models to establish biological exposure indicators. Annals of Occupational Hygiene 34: 639-651.

Fiserova-Bergerova (Thomas), V. and J. T. Pierce. 1989. Biological monitoring V: Dermal Absorption. Applied Industrial Hygiene 4:14-21.

Fiserova-Bergerova, V. 1993. Biological Monitoring VIII: Interference of alcoholic beverage consumption with biological monitoring of occupational exposure to industrial chemicals. Applied Occupational and Environmental Hygiene 8:757-760.

Industrial Health and Safety. Human Biological Monitoring of Industrial Chemicals Series. (EUR 8476 EN, 1983; EUR 8903 EN, 1984; EUR 10704 EN, 1986; EUR 11135 EN, 1987; EUR 11478 EN, 1988; EUR 12174 EN, 1989) L. Alessio, A. Berlin, R. Roi, M. Boni, eds. Commission of the European Communities, Luxembourg.

Lauwerys, R. R. 1983. Industrial Chemical Exposure: Guidelines For Biological Monitoring. Biomedical Publications, Davis, California.

Lauwerys, R. R., and P. Hoet. 1993. Industrial Chemical Exposure: Guidelines For Biological Monitoring. Lewis Publishers, Boca Raton, Florida.

Leung, H. W., and D. J. Paustenbach. 1988. Application of pharmacokinetics to derive biological exposure indexes from threshold limit values. American Industrial Hygiene Association Journal 49:445-450.

Ogata, M., V. Fiserova-Bergerova, and P. O. Droz. 1993. Biological Monitoring VII: Occupational Exposures to Mixtures of Industrial Chemicals. Applied Occupational and Environmental Hygiene 8:609-617.

Piotrowski, J. K. 1977. Exposure Tests for Organic Compounds in Industrial Toxicology. DHEW (NIOSH) Pub. No. 77-144. U.S. Govt. Printing Office, Washington, D.C.

STUDY DESIGN

Biomarkers and Occupational Health: Progress and Perspectives. 1995. Pp. 105-108

Validation of DNA Adducts as Biological Markers of Carcinogen Exposure and Effects

Frederica Perera

Molecular epidemiology uses an interdisciplinary approach, combining the most recent techniques in molecular biology with more traditional epidemiological methods in order to identify precise factors in disease causation in humans. Disease prevention is the keystone of molecular epidemiology; the goal is to develop molecular markers of disease at the preclinical stage, thus signaling the need for intervention for those at high risk before disease can occur. Efforts are ongoing to validate biological markers for carcinogens found in the ambient environment, in a number of occupational and clinical settings, and in cigarette smoke. These biologic markers, which are measured directly in human cells and tissues, can provide quantitative data on the biologically effective doses of certain environmental toxicants as well as on their early molecular effects. Biologic markers of susceptibility, including genetic/metabolic and nutritional factors, are also being evaluated. During the past year, our research efforts have resulted in substantial progress in validating biomarkers in populations with well-defined exposures. Some examples follow.

Cross-sectional studies and longitudinal studies have been conducted in "healthy" populations with well-defined exposures to carcinogens in order to ascertain the relationship between exposures and cellular molecular effects. These populations include urban residents exposed to air pollution from combustion of fossil fuel, foundry workers, and lung cancer cases and controls. These studies are providing a detailed understanding of the reproducibility of methods and the variation in molecular response among individuals having comparable exposures.

Levels of DNA Adducts with Increasing Exposure to Benzo(a)pyrene in Ambient Air

In an investigation of the effects of ambient air pollution, several biologic markers were analyzed in the blood samples from a group of 39 persons from Gliwice, a heavily industrialized region of Poland, during the winter and summer of 1990 (Perera et al., 1992). Benzo(a)pyrene (BP) levels in ambient air in this region ranged from 15 ng/m^3 in the summer to 66 ng/m^3 during the winter months. Peripheral blood samples were also taken from 49 residents of Biala Podlaska, a rural town in eastern Poland, with ambient BP levels that are approximately 10% of the levels measured in Gliwice. None of the study participants were employed in industries that generated polycyclic aromatic hydrocarbons (PAHs).

After adjusting for smoking status and age, exposure to ambient pollution was significantly related to blood levels of PAH-DNA adducts as measured by the enzyme-linked immunosorbent assay (ELISA) ($p = 0.001$), to aromatic adducts measured by the postlabelling method ($p = 0.001$), to sister chromatid exchange (SCE) frequency ($p = 0.016$), and to chromosomal aberration (CA) frequency, including gaps ($p = 0.01$). Further, there was a linear relationship between ambient BP levels and both measures of DNA adducts as well as with SCE and CA frequencies ($p < 0.01$). Finally, a correlation was found between levels of DNA adducts measured by ^{32}P-postlabelling and chromosomal aberrations, providing a molecular link from environmental exposure to genetic alterations relevant to cancer and reproductive risk (Perera et al., 1992).

Levels of DNA-PAH Adducts with Increasing Occupational Exposure to Benzo(a)pyrene

A study of Finnish foundry workers exposed to PAHs and other carcinogens in the workplace has recently completed its third year of sampling (Perera et al., 1993). Blood and urine samples were taken from 64 workers in December, 1993. To date, 158 blood and urine samples have been obtained. Workplace levels of benzo(a)pyrene ranged between 2 and 60 ng/m^3 in air. PAH-DNA adducts measured by ELISA and aromatic-DNA adducts measured by the postlabelling method increased with exposure; however, there was marked individual variation in levels of biomarkers measured among the exposed workers.

In the first year of sampling, when exposure levels were higher, HPRT mutation frequency was highly correlated with levels of PAH-DNA adducts ($r = 0.67, p = 0.004$). In the second year of sampling, levels of BP in the workplace were reduced substantially. This decline was reflected in levels of PAH-DNA and PAH-albumin adducts as well as in levels of several markers of effect. This is the first study to find an association between PAH-DNA adduct levels and gene mutation frequency in humans and is consistent with experimental data for PAHs.

Levels of DNA-PAH Adducts in Lung Cancer Tissue

In a molecular epidemiologic study of lung cancer that was initiated to validate biomarkers that might be predictive of future risk (Tang et al., 1993), PAH-DNA adducts were measured by ELISA in peripheral blood leukocytes collected prior to or at surgery. Information on smoking, diet, and occupational exposure was obtained. The levels of PAH-DNA adducts in the blood of lung cancer cases and controls were evaluated to determine whether, when exposures to PAHs and other mutagens/carcinogens are comparable, the lung cancer cases have higher levels of PAH-DNA binding. Such a finding would suggest that the ability to efficiently activate and bind carcinogens may have been a risk factor in their disease. A total of 136 lung cancer patients at the Columbia Presbyterian Medical Center in New York City have participated in this study. Controls consisted of 115 patients with no prior history of cancer or benign lung disease who were being treated at the same hospital for orthopedic conditions. A detailed questionnaire was administered to both cases and controls by a trained interviewer. Lung tissue (tumor and/non-tumor) was obtained from 35 of the lung cancer patients.

Using multivariate linear regression adjusting for age, gender, season, ethnicity, and the number of cigarettes smoked per day, PAH-DNA adduct levels in white blood cells were significantly associated with lung cancer ($p = 0.01$) when PAH-DNA adduct levels were stratified into high and low categories. PAH-DNA adducts in leukocytes were significantly higher in smokers and ex-smokers than in non-smokers among cases alone and among cases and controls combined. However, in the control group, the only significant difference in adduct levels was between current smokers and non-smokers. A dose-response was observed among the 51 cases who were current smokers; linear regression analysis showed higher PAH-DNA adduct levels with heavier smoking as measured in cigarettes per day. When the same analysis was performed on 22 current smokers in the control group, there was no association between PAH-DNA adduct levels and amount of smoking. Levels of DNA adducts in lung tumor tissue were higher in current smokers than in former smokers and non-smokers (Tang et al., 1993).

A nested case control study of lung cancer is ongoing to investigate the predictive value of biomarkers, i.e., the extent to which the markers present in blood samples drawn while individuals are healthy will be useful in identifying future lung cancer cases.

CONCLUSIONS

In summary, the formation of PAH-DNA adducts and related biomarkers has been demonstrated to be sensitive to a variety of low-level environmental exposures. In addition, there is an association between lung cancer risk and PAH-DNA adduct formation after controlling for the amount of smoking. These findings suggest that adducts can serve not only as an environmentally relevant dosimeter, but that they may also indicate heightened risk of cancer. These results indicate the potential usefulness of biomarkers as "early warning systems" to guide intervention strategies. However, careful validation is needed before biomarkers can be applied to population "screening."

REFERENCES

Perera, F., K. Hemminki, E. Grzybowska, G. Motykiewicz, J. Michalska, R. Santella, T.L. Young, C. Dickey, P. Brandt-Rauf, I. DeVivo, W. Blaner, W-Y Tsai, and M. Chorazy. 1992. Molecular and genetic damage from environmental pollution in Poland. Nature 360:256-258.

Perera, F., D. Tang, P. O'Neill, W. Bigbee, R. Albertini, R. Santella, R. Ottman, W-Y Tsai, C. Dickey, L. Mooney, K. Savela, and K. Hemminki. 1993. HPRT and Glycophorin A mutations in foundry workers: Relationship to PAH exposure and to PAH-DNA adducts. Carcinogenesis 14:969-973.

Tang, D. L., R. M. Santella, M. A. Blackwood, D. Warburton, J. Luo, T. L. Young, J. Mayer, W. Tsai, and F. P. Perera. 1993. A case-control molecular epidemiology study of lung cancer. American Association of Cancer Research, Proceedings.

Biomarkers and Occupational Health: Progress and Perspectives. 1995. Pp. 109-115

The Development, Validation, and Application of Biomarkers for Early Biologic Effects

Nathaniel Rothman, Walter F. Stewart, Charles S. Rabkin, and Paul A. Schulte

Traditionally, epidemiologic studies have focused on observable disease as the outcome of concern. This approach may not allow for the timely assessment of potential carcinogenic hazards, since cancer is often characterized by long induction and latency periods of 20 or more years. Thus, risk information obtained from epidemiologic studies may no longer be relevant because current occupational exposures have changed in either degree or type.

One potential solution to this dilemma is to use biomarkers of early biologic effect that are valid predictors of disease risk. Using a biomarker outcome, instead of clinical disease, may allow for identification of occupational risk factors earlier in the exposure-disease sequence at a time when control measures can be implemented.

A wide spectrum of biomarkers may be applicable for screening cancer risks. These include measures of chromosome damage such as chromosome aberrations, micronuclei, and sister chromatid exchange, and markers of somatic cell mutation, such as the HPRT, HLA and glycophorin A assays (Compton et al., 1991). Further, the polymerase chain reaction and other technologies may soon spawn a new generation of exceptionally sensitive and specific assays for DNA damage. However, the relationship linking the external exposure, the biomarker, and disease risk must be understood before the biomarker can be used to evaluate workplace hazards. Typically, a series of steps must be completed before a candidate biomarker assay is adequately validated. Laboratory and population-based studies contribute to the development and application of a biomarker useful in public health practice. These studies may be categorized into laboratory studies, transitional studies, etiologic studies, and public health applications (Schulte et al., 1993; Schulte, 1993). We have developed a systematic approach that identifies the steps necessary for evaluating a potential biomarker (Rothman et al., 1994). In this paper, we present the component relevant for the validation of biomarkers of early biologic effect (Table 1).

TABLE 1. STAGES IN VALIDATING BIOMARKERS
OF EARLY BIOLOGIC EFFECTS

STUDY DESIGN	STUDY OBJECTIVE
I. LABORATORY STUDIES	Develop assay and test exposure-marker relationship in animals/*in vitro* systems.
II. TRANSITIONAL STUDIES	
Developmental	Assess sample collection and processing needs; evaluate assay reliability and accuracy.
Characterization	Evaluate prevalence and determinants of marker in general population.
Applied	Evaluate exposure-marker relationship in select populations.
III. ETIOLOGIC STUDIES	
Case-control	Evaluate marker-disease relationship for markers that are fixed and unaffected by disease.
Prospective cohort	Evaluate marker-disease relationship for markers that are transient and/or affected by disease.
IV. APPLICATIONS	
Risk Assessment	Identify high risk occupational settings.
Prevention	Clinical use of marker to screen and/or treat individuals.

LABORATORY STUDIES

Biomarker assays are developed in the laboratory. Historically, candidate biomarkers for population-based studies emerged as a by-product of basic research in carcinogenesis. More recently, the development of biomarkers has been an explicit goal of laboratory research. Experiments in animals allow for controlled observations of the relationship between exposure, biomarker formation, and disease. However, this relationship must be evaluated in humans before a biomarker can be used in population screening.

TRANSITIONAL STUDIES

The initial steps in the evaluation of biomarkers in human populations are performed in transitional studies (Hulka, 1991; Hulka and Margolin, 1992) which serve to "bridge the gap" between the development of markers in the laboratory and their eventual application to studies of human disease. They are performed on generally healthy subjects and evaluate such issues as assay reliability and the exposure-marker relationship. Transitional studies can be further classified into developmental, biomarker characterization, and applied studies.

Developmental Studies

When a promising new biomarker emerges from the laboratory, the reliability and accuracy of its assay are initial concerns. However, accuracy may be elusive when there is no "gold standard" (e.g., for cytogenetic and mutation assays). If assay reliability is acceptable, the procedures for the collecting, processing, shipping, and storing of biological specimens must then be optimized and adapted for field conditions. Often, minor variations in the handling of biologic specimens can substantially alter assay results. Thus, developmental studies are an iterative process in which both sample acquisition and assay conditions are optimized.

Characterization Studies

Characterization studies evaluate the presence of a new marker in the general population, identifying factors such as age, sex, or smoking patterns that may influence a marker. These studies collect information that allows subsequent investigations to be designed more effectively. For example, control populations can be matched to exposed groups on identified confounders, and the required sample size can be estimated.

Applied Studies

Applied studies evaluate healthy subjects who are exposed to the chemical or physical agent of interest. These studies determine the relationship between exposure and biomarker levels and evaluate the kinetics of biomarker formation and persistence. However, such studies by themselves cannot establish an association between the biomarker and an increase in disease risk.

ETIOLOGIC STUDIES

The association between a biomarker and disease risk must be established through etiologic studies before a biomarker may be used for population screening. Etiologic studies are distinct from transitional studies in that the outcome of concern is disease rather than biomarker formation. Both case-control and prospective cohort study designs may be used.

The case-control method, which compares the distribution of a marker in subjects with disease ("cases") and in subjects without disease ("controls"), is useful for the initial evaluation of some markers of early biologic effect, especially those that persist over time and are unlikely to be affected by disease status. However, the results of such studies must generally be confirmed in prospective studies in order to definitively avoid the problem of reverse causality (i.e., the marker may result from the disease, rather than be associated with risk of developing the disease).

In the prospective cohort study design, biologic samples are collected from exposed subjects (the "cohort") while they are healthy. The cohort is followed forward in time and subjects who develop disease are identified. Often, a "nested case-control" approach is used in which specimens are analyzed from cases and a sample of the cohort without disease; this approach may considerably reduce the laboratory requirements.

As discussed by Schatzkin et al. (1990), a biomarker is validated by determining its sensitivity (S), relative risk (RR), and attributable proportion (AP). In this context, sensitivity refers to the proportion of cancer cases positive for the biomarker and relative risk refers to the ratio of cancer incidence for biomarker-positive versus biomarker-negative subjects. The attributable proportion, which is that fraction of disease associated with the marker, is calculated as:

$$AP = S \times (1 - \frac{1}{RR}) \qquad (Equation\ 1)$$

(Schatzkin et al., 1990). The attributable proportion in Equation 1 varies from 0 to 1.0, with 1.0 representing universal association and 0 representing lack of association. The same calculations apply regardless of whether a biomarker reflects processes directly on the causal pathway from exposure to disease, or is a surrogate for such processes (Steenland et al., 1993).

Several biomarkers relevant to chronic disease risk have been successfully validated in prospective cohort studies and are being extensively applied in the public health setting. Examples include elevated serum cholesterol for risk of atherosclerotic heart disease and HIV seropositivity for risk of AIDS. The validation of biomarkers for chemical or physical carcinogens is less well established, but several promising measures have emerged. For example, chromosomal aberrations have been associated with an increased risk of cancer (all sites combined) in two small prospective cohort studies (Hagmar et al., 1994; Bonassi et al., 1994). Notably, the former study (Hagmar et al., 1994) found that sister chromatid exchange was not associated with excess cancer risk. Thus, research is in progress to sift through candidate biomarkers to identify which ones are relevant for disease risk.

Finally, it is possible that a biomarker will be associated with cancer risk only in populations with particular exposure patterns. Populations with differing sets of exposures will need to be studied in order to determine the generalizability of a particular biomarker-disease relationship.

PUBLIC HEALTH APPLICATIONS

A validated biomarker may be used to study potential disease risk in workers with particular past or current exposure patterns. In general, applicability of a biomarker at the group level (risk assessment) will precede that at the individual level (clinical prevention) because of logistical and interpretive considerations. For example, an assay may be prohibitively expensive, detection will generally imply only a given probability of future disease (which may be difficult to precisely calculate for an individual), or corrective measures (apart from avoidance of known exposures) may be undefined. However, these limitations should not impede the use of a biomarker for screening select populations, so long as participants are informed in advance of the uncertain implications of test results at the individual level.

Screening working populations with validated biomarkers has the potential to identify hazardous work environments early enough to allow for timely intervention. The task at hand is to characterize a biomarker's relationship with both exposure and disease so that it may become useful in practical public health applications.

REFERENCES

Bonassi, S., P. Padovani, C. Lando, D. Vecchio, S. Parodi, and R. Putoni. 1994. Cohort study on chromosome aberration and cancer mortality (Abstract). Proceedings of the American Association for Cancer Research 35:292.

Compton, P. J. E., K. Hooper, and M. T. Smith. 1991. Human somatic mutation assays as biomarkers of carcinogenesis. Environmental Health Perspectives 94:135-141.

Hagmar, L., A. Brogger, I-L. Hansteen, S. Heim, B. Högstedt, L. Knudsen, B. Lambert, K. Linnainmaa, F. Mitelman, I. Nordenson, C. Reuterwall, S. Salomaa, S. Skerfuing, and M. Sorsa. 1994. Cancer risk in humans predicted by increased levels of chromosomal aberrations in lymphocytes: Nordic study group on the health risk of chromosome damage. Cancer Research 54:2919-2922.

Hulka, B. S. 1991. Epidemiological studies using biological markers: Issues for epidemiologists. Cancer Epidemiology, Biomarkers and Prevention 1:13-19.

Hulka, B. S., and B. H. Margolin. 1992. Methodological issues in epidemiologic studies using biologic markers. American Journal of Epidemiology 135:200-209.

Rothman, N., S. Stewart, and P. A. Schulte. 1994. A matrix of biomarker and study design categories in cancer epidemiology (submitted).

Schatzkin, A., L. S. Freedman, M. H. Schiffman, and A. M. Dawsey. 1990. Validation of intermediate end points in cancer research. Journal of the National Cancer Institute 82:1746-1752.

Schulte, P. A. 1993. Use of biological markers in occupational health research and practice. Journal of Toxicology and Environmental Health 40:359-366.

Schulte, P. A., N. Rothman, and D. Schottenfeld. 1993. Design considerations in molecular epidemiology. Pp. 159-198 in: Molecular Epidemiology: Principles and Practices, P. A. Schulte, and F. P. Perera, eds. Baltimore, Williams and Wilkins Co.

Steenland, K., J. Tucker, and A. Salvan. 1993. Problems in assessing the relative predictive value of internal markers versus external exposure in chronic disease epidemiology. Cancer Epidemiology, Biomarkers and Prevention 2:487-491.

Biomarkers and Occupational Health: Progress and Perspectives. 1995. Pp. 116-119

The Development, Validation, and Application of Biomarkers for HIV

Charles S. Rabkin and Nathaniel Rothman

Human Immunodeficiency Virus (HIV) seropositivity is a paradigm of the ideal marker of biologically effective dose; it is associated with both exposure to the HIV virus and is nearly universally linked with the development of AIDS and related health effects. Someone who is HIV-positive is likely to develop AIDS, whereas without HIV-positivity, AIDS is unlikely. But this (albeit over-simplified) scheme has only become recognized over time. The history of the development, validation, and application of assays for HIV exposure portrays a process that, to a certain extent, can be generalized to other markers that may have application in screening populations exposed to chemical and physical agents. An overview of this process and the nomenclature describing it has been presented previously in this volume. In this paper, HIV seropositivity is used as an example of this approach.

The HIV Chronology

After HIV was isolated, the first step in generating a biomarker for HIV infection was the development in the *laboratory* of a test for antibody to the virus using an enzyme-linked immunoassay. These tests utilize the specificity of antibody-antigen reactions. *Transitional studies*, defined as the initial application of biomarkers in select or general population groups, were used to evaluate the utility of this assay. These include *developmental, characterization* and *applied studies*. *Developmental studies*, which evaluate aspects such as assay sensitivity and specificity, demonstrated that the enzyme-linked immunoassay had a problem with cross-reactivity, i.e., there were instances when subjects had antibodies directed to something else, but that cross-reacted with HIV. This led to the development of a new test, the Western blot, which simultaneously assessed antibodies to multiple antigenic determinants. This test diminishes, but does not eliminate, the problem of non-specificity of the enzyme immunoassay.

With these refined tools, it was then possible to perform *characterization studies*, which determine the prevalence of the marker in the general population or in selected populations with specific exposure or disease-risk characteristics. Studies of HIV seropositivity in several populations documented elevated prevalence of HIV antibodies in groups at higher risk of AIDS, such as men who engage in homosexual activities, intravenous drug users, and transfusion recipients (Weiss et al., 1985).

Subsequently, *applied transitional studies* were carried out. These are studies of healthy individuals in whom the biomarker is the outcome variable. Although at this stage of research HIV assays had not yet been shown to predict an increased risk of AIDS itself, these assays were useful for gaining insight into the exposure-biomarker relationship. These studies determined that within groups at higher risk of HIV, HIV seropositivity was quantitatively associated with AIDS risk behaviors or exposures including the number of sex partners, duration of intravenous drug use, and the amount of transfused blood products (Goedert et al., 1984).

With *etiologic studies*, the biomarker's association with disease was directly investigated. Initial *case-control studies* showed an unexpectedly stronger association of antibody reactivity with AIDS-related complex than with AIDS (Gallo et al., 1984). This paradox was subsequently resolved by *prospective studies* which demonstrated that ill persons progressing to AIDS can lose antibody to HIV. Prospective studies demonstrated that antibody reactivity always preceded AIDS (Jaffe et al., 1985). In this way, HIV antibody was validated as a biomarker for AIDS risk. Then, *intervention studies* could use HIV seropositivity rather than AIDS as the outcome to assess HIV preventive measures such as the use of condoms, counselling, and needle exchange (van Griensven et al., 1989).

This base of knowledge then allowed for public health *application* of these tests. HIV serotesting can be used for *risk assessment* to identify high-risk populations and occupational settings. It can also be used for *prevention* programs (as opposed to prevention research) in screening and early intervention in individuals. Currently, HIV seropositivity is a validated biomarker that has widespread application.

CONCLUSIONS

The experience with HIV testing has deeply affected medical practice and societal attitudes. HIV testing has stimulated enormous attention to the issues of confidentiality, insurability, employability, informed consent, individual rights, and other concerns. HIV thus serves as an example and a forerunner for development and application of biomarkers for chemical and physical exposures.

TABLE 1. STAGES IN THE VALIDATION OF HIV ANTIBODY ASSAYS

STUDY DESIGN	STUDY GOALS AND FINDINGS
I. LABORATORY STUDIES	Development of ELISA for HIV.
II. TRANSITIONAL STUDIES	
Developmental	Need realized for more specific, confirmatory Western blot.
Characterization	Elevated prevalence in AIDS risk groups (Weiss et al., 1985).
Applied	Quantitative association with AIDS risk factors (Goedert et al., 1984).
III. ETIOLOGIC STUDIES	
Case-Control	Unexpected stronger association with AIDS-related complex than with AIDS (Gallo et al., 1984).
Prospective Cohort	Demonstration that HIV seropositivity predicts risk of AIDS (Jaffe et al., 1985).
Experimental (Intervention)	Assessment of AIDS preventive measures by effect on HIV seropositivity (van Griensven et al., 1989).
IV. APPLICATIONS	
Risk Assessment	Identifying high risk populations and occupational settings; projecting future burden of disease.
Prevention	Screening and early intervention.

Table 1 outlines the development, validation and application process using HIV serologic testing. After more than 12 years of work, the laboratory test for antibody to HIV has been enhanced, characterized, and applied in healthy populations, and its relationship to AIDS has been demonstrated in case-control, cohort, and finally intervention studies. Today, this test has important applications in risk assessment and intervention.

Biomarkers for chemical and physical exposures are less well developed, but HIV is a useful model for how a biomarker may be validated. Systematic validation of a biomarker is essential to advance knowledge of its implications and ensure a rational and appropriate approach to its practical applications.

REFERENCES

Gallo, R. C., S. Z. Salahuddin, M. Popovic, G. M. Shearer, M. Kaplan, B. F. Haynes, T. J. Palker, R. Redfield, J. Oleske, and B. Safai. 1984. Frequent detection and isolation of cytopathic retroviruses (HTLV-III) from patients with AIDS and at risk for AIDS. Science 224:500-503.

Goedert, J. J., M. G. Sarngadharan, R. J. Biggar, S. H. Weiss, D. M. Winn, R. J. Grossman, M. H. Greene, A. J. Bodner, D. L. Mann, and D. M. Strong. 1984. Determinants of retrovirus (HTLV-III) antibody and immunodeficiency conditions in homosexual men. Lancet 2 (8405):711-716.

Jaffe, H. W., W. W. Darrow, D. F. Echenberg, P. M. O'Malley, J. Getchell, V. S. Kalyanaraman, R. H. Byers, D. P. Drennan, E. H. Braff, and J.W. Curran. 1985. The acquired immunodeficiency syndrome in a cohort of homosexual men. A six-year follow-up study. Annals of Internal Medicine 103:210-214.

van Griensven, G. J. P., E. M. M. de Vroome, J. Goudsmit, and R. A. Coutinho. 1989. Changes in sexual behavior and the fall in incidence of HIV infection among homosexual men. British Medical Journal 298:218-221.

Weiss, S. H., J. J. Goedert, M. G. Sarngadharan, A. J. Bodner, R. C. Gallo, and W. A. Blattner. 1985. Screening test for HTLV-III (AIDS-agent) antibody: Specificity, sensitivity, and applications. Journal of the American Medical Association 253:221-225.

Biomarkers and Occupational Health: Progress and Perspectives. 1995. Pp. 120-130

Quantitative Decision Support Systems for Surveillance and Clinical Applications

Philip Harber

Quantitative decision models are formal methods of dealing with uncertainties and complex questions. These approaches are particularly relevant to situations in which the number of variables is large and for which there is incomplete information such as in the decision as to whether or not to incorporate biomarkers into a medical surveillance program.

Many quantitative, arithmetic models are available and are potentially useful in simulating and simplifying complex situations. Qualitatively, they force clear specification of the factors involved and provide a *process* for resolving conflicts of opinion. Quantitatively, the calculations themselves may provide useful and accurate conclusions.

In assessing the utility of testing procedures, it is not adequate to rely on traditional measures such as the sensitivity and predictive value of the test procedure itself. Rather, the net benefit of a program including the test procedure must be evaluated. That is, results of using the procedure, rather than its innate characteristics, should dictate whether it should be used.

The approaches used are not epidemiologic, although they may utilize epidemiologic data. They are particularly applicable to discussions of the application of biomarkers in medical surveillance because of the complexity of the issues, the differing fundamental "biases" of participants in discussions, and the need to make clear policy and rationally explicable decisions even in the absence of complete empiric data.

These concepts may be illustrated with some examples of work we have done.

Pre-Placement Medical Evaluation

The first example deals with implications of information that biomarker technology will yield about *individuals*. If a biomarker test (e.g.,

120

HLA phenotype) shows that an individual was at high risk for a disease associated with a certain exposure, should that person be placed in a job with the potential for that exposure? Compliance with the Americans with Disabilities Act (ADA) as well as "philosophic" differences in the proper balance between right to work and worker protection raise difficult questions. Facing such decisions, we developed a quantitative model to help make adequate choices (Harber et al., 1993).

The use of a quantitative decision model requires explicit enumeration of the factors to be considered. For example, if there is a question of whether HLA allele type should be used to determine whether someone should be classified as a beryllium exposed worker (see chapters by Saltini and by Rossman), it is necessary to determine what empiric data should affect the decision. This approach allows separation of the actual data from the inferences. Thus, if data change, the process does not need to be reexamined.

Approaching the placement decision problem generically, a minimum set of factors is necessary. These factors fall into two groups: factual information and expressions of societal values.

The first consideration is the overall likelihood that an adverse outcome will occur. How common is beryllium disease in a population?

The second factor is *imminence*. Clearly, many people feel, and the ADA implicitly states, that something that happens soon is more important than something that will occur in the distant future. We therefore considered temporal inhomogeneity (how the likelihood of the adverse event is distributed over time).

The third factor is *personal risk modifiers*. While general likelihood refers to the average risk in the worker population or in a particular job or industry, personal risk modifiers lower or increase an individual's risk compared to the average person's likelihood. Either his or her job is unique or his or her personal characteristics are unique (e.g., a specific HLA allele type may greatly modify the risk of chronic beryllium disease). Of course, these risk modifiers may also have an effect on an outcome. Some may be important early and some gain importance later.

The next factor is the *severity of the outcome*. Leukemia is worse than contact dermatitis. For some outcomes, there are differences in severity, some persons may be more severely affected than others.

All of the above factors are empirically knowable. However, in most situations not all this information is available from epidemiologic studies. Often one must rely upon simple case studies or just intuition; but at least theoretically, they are empirically knowable.

In contrast, there are also factors that go into the overall decision process that are not empirically knowable. Even given infinite money and infinite time, it is not possible to do a study to empirically determine the answer to "value" questions.

The first value factor is the determination of relative importance of effects upon self (personal illness/injury), others (public health), or productivity. Which carries more weight: an outcome that adversely affects the person or one that hurts the public health? There is no way that can be "known"; it is a societal value. A societal value in the workplace setting is what is referred to as the future discount. This is a classic decision analytic and modeling concept. The weighing that should be given to the future discount is not empirically knowable.

Clear delineation of "value" from empiric factors will significantly clarify discussion. Often, apparent disagreements can be resolved if broken into small parts (i.e., "atomic" approaches).

Once the factors are explicitly identified, they are combined in a quantitative model to determine the impact estimate. The overall impact estimate is the probability that the event will occur, times the personal modifiers, times the severity, times the future discount.

The decision modeling approach helps define what data are needed to resolve differences of opinion or to set policy. Where information is known, empiric parameter estimates may be used. When they are not known, decisions may be delayed until empiric research may be conducted or at least until expert opinion can be obtained.

The advantages of developing even simple models include:

- The method facilitates integrated decision-making. In other words, it is necessary to consider all the factors, not simply to focus attention on one aspect (e.g., the severity of the disease).

- These models can clinically utilize epidemiologic data where available and define the need for additional data. It is thus a method for prioritizing research efforts. For example, if some parameters are completely unknown, there is no point in measuring a known parameter to 0.01 precision. If the value of an intervention is unknown, extreme epidemiologic precision in incidence rates may not be helpful.

- The method fosters specification of the biases, the utilities, and values.

- The method permits individualized algorithms. One can avoid "class decisions" by which all people in some job or exposure category are treated in the same fashion. One can easily adjust for personal factors in an explicit and consistent fashion, making charges of discrimination less likely.

- Such models force reconsideration of certain classic epidemiologic screening theory definitions. For example, "predictive value" in cancer screening usually relates to the ability of a simple test to predict the current presence of a cancer. This is very different from the term "predictive value" when applied to a biomarker to predict an event that may happen 30 years from now.

In addition, when dealing with non-cancer outcomes, terms such as "sensitivity" and "predictive value" often really mean "concordance." Classic use of these terms requires the existence of a "gold standard" for reference which is independent of the test itself.

How Does Epidemiology Differ from Decision Support Systems?

Decision support is based largely on epidemiology, but goes beyond the limits of epidemiology. Important differences between epidemiology and decision support include the following:

- Epidemiology is empiric and data based, whereas decision support can actually use opinion where there are no data. Epidemiologic data are not available for many of the factors that are described above.

- Values are nominally excluded from epidemiologic studies, whereas values are explicitly included in decision support systems. Thus, in decision support systems, differences of opinion about values need not masquerade as empiric science.

- Epidemiology generally works with a relatively small number of variables despite multivariate analytic methods.

- Epidemiology generally focuses upon using data to test hypotheses, limiting type II (false negative) errors, whereas decision support is a likelihood maximization method, seeking the best answer despite poor or incomplete data.

- Epidemiology tends to be deductive, while decision support is often inferential.

- Often epidemiologic studies are very expensive and require considerable time in order to design, conduct, and analyze data resulting from a study.

- Epidemiology is nominally definitive. When a study is completed, one assumes that the answer represents "truth." Decision support, however, is basically a collection of methods to foster discussion.

Thus, while epidemiologic data are still the optimal basis for making very important decisions, there are times when epidemiology is not adequate because the information is not available or would be too costly to obtain, or because it would take decades to reach conclusions. Given the pressure to develop rational approaches to biomarker use *today*, the delays and costs of epidemiologic research inherently preclude sole reliance upon epidemiology. For example, the very large NCI Lung Cancer cytology screening project did not yield an unequivocal answer about the utility of such screening.

Models For Screening Program Optimization

Often, screening or other surveillance testing needs to be done on a recurring, periodic basis. Although selection of the optimum interval can have a major impact upon overall efficacy and cost, empiric trials with different intervals are not feasible. If a model of disease development and progression is developed, data from a single interval implementation can be used to empirically estimate model parameters for testing at other intervals. Classic screening calculations (e.g., sensitivity) are based upon a single test application (i.e., a "prevalence screen"), whereas most applications are periodic.

Johns Hopkins University participated in a large study of sputum cytology screening. Approximately 10,000 men were randomized into two groups and followed for five years with and without cytology testing at four-month intervals. The question posed for model development was, "based upon the results of screening three times per year, what would be the yield if testing were done at other frequencies (e.g., every two years or every month)?"

The four-month interval study cost $25 million to conduct, and therefore, studies of other intervals would not be feasible. In order to examine the benefits of screening at different time intervals, we developed a simple, three-stage stochastic model (Parker et al., 1983). The three states were: n1 = healthy; n2 = early lung cancer (treatable and detectable by screening); n3 = advanced (incurable). The model itself is a simple Markov model which assumes that at any point in time there is a fixed transition probability to go between any two states. There are state transition probabilities from n1 to n2 (T_{n1-n2}), n2 to n3 (T_{n2-n3}), and n1 to n3 (i.e., some people go directly from healthy to incurable without becoming part of the detectable/curable stage).

The transition parameters were estimated based on the empiric data from the Johns Hopkins study which used the four-month interval screen. Once developed, the model allowed determination of the effect of different intervals. As the screening interval either increases or decreases, the yield varies.

Program Implementation: Public Health Perspective

The first example illustrated application to a clinical problem, one that was focused on a decision affecting a single individual. The model described here is a public health approach, providing the framework for the choice of a program for a population group (Harber and Hsu, 1991). Program development from the public health perspective requires compromises due to resource limitations and, equally importantly, because of conflicting goals.

The *process* of model development requires explicit definition of the information needed and forces an explicit statement of goals. In particular, implementation methods must be guided by the net benefit on human health of any potential preventive actions that may be triggered by the test results.

Therefore, aspects such as the ability to identify the at-risk population or effectively communicate results to workers may be critically important. It has been suggested that in the near future, a flow cytometry device in a doctor's office may be used to predict future susceptibilities to disease 30 years hence. Although the ability to predict the risk of future disease may be improving dramatically, any such advances cannot always be directly translated into useful preventive interventions. For example, clinicians have had limited success in helping patients stop smoking cigarettes, even though this is a well known personal risk factor. Another example is the national coalworkers surveillance program. Miners are radiographically examined on a regular basis, and if early pneumoconiosis is found, that worker is offered transfer to a non-dusty job with wage retention. There are good epidemiologic data showing that this transfer will prevent progressive massive fibrosis (the disabling form of coal dust disease). Despite this, only a minority of workers who are detected as positive have actually accepted the transfer option (Hoffman, 1986).

Table 1 presents the specific factors to be considered in evaluating whether to establish a medical surveillance program. As in the first example, some of these factors are potentially empirically knowable (e.g., the disease incidence), whereas others are value-based (e.g., discounting for future effects). Some of the factors are dependent on the test itself, whereas many others relate to the *system* in which the test will be used. For example, advances in the technological base for biomarker testing (e.g., specimen processing throughputs) will make access to the proper target population a limiting factor in surveillance applications.

The metric for assessment of program outcomes must be defined *a priori*. When choices are present, is the total cost or the cost-benefit ratio the determining factor? To what extent should long-term benefit (or cost) be discounted? Should cost be referenced to individual cases or to the total population?

Prevention efforts can be characterized as primary, secondary, or tertiary; they can be further characterized by their focus on the person, the environment, or upon administrative systems. While it is traditional in public health practice to emphasize primary prevention, this may not always be the most effective use of resources. For example, application of a simple model that we developed showed that for smoking, primary prevention is most effective. However, the analysis for occupational asthma indicates (rather counter-intuitively) that the best approach, according to the parameters utilized, is secondary prevention.

Other Methods

Numerous other decision support systems are available. These systems use artificial intelligence applications to provide expertise within a limited domain. They are particularly applicable when the range of possible combinations of factors is so large that a single model could not be developed (Harber and McCoy, 1989: Harber et al., 1991a, 1991b, 1994). Many systems are relatively simple to modify. General components of decision support systems include the following:

- Data Requirements
- Inference system that is independent of data components but robust to differences in quantity of data
- A system to manage uncertainties
- Incorporation of biases

TABLE 1. FACTORS AFFECTING CHOICE OF A MEDICAL SURVEILLANCE PROGRAM

DISEASE CHARACTERISTICS

Attack Rate	Percent of workers who will get the disease
Latency	Mean time (years) from exposure to disease
Disease Cost	"Cost" of disease in relative units

CAPABILITIES

Access to Group/Site	Ability to conduct program onsite with cooperation
Industrial Hygiene Availability	Industrial hygiene/engineering availability
Medical Availability	Medical support availability (0-100 percent)
Occupational Medicine	Specifically qualified Occupational Medicine staff

WORK SITE CHARACTERISTICS

Workforce Size	Total number of workers
Discount Rate	Percent annual "discount" of value of prevention now for benefit later

INTERVENTIONS

Percent Efficacy	Percent of cases prevented if worker is in target group
Prevention Cost/Worker	Cost/worker in target group
Permanence	How long the intervention lasts
Delay	Delay (years) between intervention and benefit
Percent Targeted	Percent of workers in the target population
Target Specificity	Percent of cases which are included in the target population

NEEDS OF INTERVENTION

Industrial Hygiene/ Engineering	Requirement for industrial hygiene expertise
Clinical Intervention	Requirement for clinical expertise
Occupational Medicine	Requirement for occupational medicine expertise
Group/Site Access	Requirement for access to worksite

Neural network models are particularly adept at pattern recognition. They may be applicable when there are complex data structures.

At the level of the individual, neural models may recognize patterns suggestive of illness. On a group basis, neural models may allow detection of patterns of exposure and disease without the constraints of traditional hypothesis-testing statistical methodology.

Object-oriented databases (OODBs) also hold promise. Currently, environmental and human exposure groups are defined by relatively simple schemata, typically hierarchical data structures. Occasionally, a relational model is employed. Both of these approaches, however, require the assumption of homogeneous exposure groups (HEGs). All workers within an HEG are considered to have similar exposure patterns and require similar preventive interventions (e.g., screening). Studies in industrial settings have demonstrated that in actuality, HEGs are rarely homogeneous. More sophisticated data systems (e.g., OODBs) hold the promise of facilitating more person- and site-specificity in describing exposures and selecting interventions. Within commonly used database technology, such individualization often converts data to anecdotes.

Classic decision analysis methods are also highly applicable to a variety of situations. Decision trees are constructed to map possible outcomes and calculate relative probabilities and utilities of outcomes as a function of choices (Harber and Rappaport, 1985; Harber, 1985; Auerbach and Harber, 1989). Sensitivity analysis determines whether a decision is actually sensitive to (affected by) the value of an unknown parameter. For example, if a public health program should be instituted whether disease incidence is $1/10^6$ or $10/10^6$ person-years, there is little need to conduct studies to carefully define the rate. Receiver operating characteristic curves can define the optimum cutpoint for initiating an action when test results are a continuous variable. For example, we showed that for occupational asthma, the impact of the intervention, not the characteristics of the spirometry test procedure itself, should define the value constituting a positive test.

Implications For Surveillance and Clinical Applications of Biomarkers

There are several reasons why the other decision support methods described above are particularly relevant to the topic of biomarkers; these include the following:

- Multiple goals
- Unstated "biases" and values

- Necessity of justifying conclusions
- Need to separate public health policy from purely political goals
- Incomplete empiric information
- Need to act quickly
- Need to define future data needs

There are differing opinions about goals and priorities; often, these are not clearly stated. Consensus building can be facilitated by clearly delineating the limited areas of difference from major areas of concurrence. For example, the criteria for selecting optimal programs can differ depending upon the goal selected: testing for immediate benefit versus testing for research purposes; human testing for case detection versus human testing as an indicator of environmental exposures (biologic monitoring). In addition, the use of a model, even if only as a learning device, can facilitate explanation of the basis for program plans. Thus, this approach can help clearly separate political/social considerations from medical/scientific concerns.

In addition, there is a mandate to establish public health programs and even in the absence of complete information biomarkers can prove useful. It is unlikely that existing epidemiologic studies alone can answer the fundamental questions needed to design programs. Furthermore, the cost and time delays would preclude depending upon future observational studies and clinical trials. Anecdotal information is similarly limited in extent and generalizability. Decision support methodology explicitly recognizes the need for action under uncertainty and can provide tools to facilitate such decisions.

REFERENCES

Auerbach, D. M., and P. Harber. 1989. Diagnostic approach to acquired immunodeficient patients with pulmonary symptoms: a decision analytic strategy. Seminars in Respiratory Medicine 10:252-257.

Harber, P. 1985. Value based interpretation of pulmonary function tests. Chest 88:874-877.

Harber, P., and S. Rappaport. 1985. Clinical decision analysis in occupational medicine: choosing the optimal FEV_1 criterion for diagnosing occupational asthma. Journal of Occupational Medicine 27:651-658.

Harber, P., and J. M. McCoy. 1989. Predicate calculus, artificial intelligence, and workers' compensation. Journal of Occupational Medicine 31:484-489.

Harber, P., and P. Hsu. 1991. Program optimization: a semi-quantitative approach. Occupational Medicine: State of the Art Review 6:145-152.

Harber, P., J. M. McCoy, K. Howard, D. Greer, and J. Luo. 1991a. Artificial intelligence assisted occupational lung disease diagnosis. Chest 100:340-346.

Harber, P., J. M. McCoy, S. Shimozaki, P. Coffman, and K. Bailey. 1991b. The structure of expert diagnostic knowledge in occupational medicine. American Journal of Industrial Medicine 19:109-120.

Harber, P., K. Czisny, P. Hsu, and J. Beck. 1994. An expert system based preventive medicine examination adviser. Journal of Occupational Medical Citation (submitted).

Harber, P., P. Hsu, and J. Fedoruk. 1993. Personal risk assessment under the Americans with Disabilities Act: a decision analysis approach. Journal of Occupational Medicine 35:1000-1010.

Hoffman, J. M. 1986. X-ray surveillance and miner transfer programs — Efforts to prevent progression of coal workers' pneumoconiosis. Annals of the American Conference of Governmental Industrial Hygienists 14:293-297.

Parker, R. D., P. Harber and L. G. Kessler. 1983. Evaluation of screening effectiveness. Journal of Medical Systems 7:11-24.

CLEANUP WORKERS AND OTHER
MEDICAL NEEDS

Biomarkers and Occupational Health: Progress and Perspectives. 1995. Pp. 133-139

Medical Surveillance at a Hazardous Waste Site

Thomas Henn

Several complicated health issues are common across hazardous waste sites. These include the difficulty of estimating potential exposure to mixed wastes, obtaining complete information on exposure by contractors, and estimating individual exposures. Generally, the individual work sites and, therefore, the agents to which the workers are exposed are uncharacterized. This does not mean that nothing is known about these sites; it means that there is incomplete knowledge about these sites, and this is a major impediment in the design of an adequate surveillance program.

This paper presents a review of the medical surveillance program which is under way at one of the United States nuclear waste sites. The program relies mainly on detailed exposure assessments and on the extensive use of protective equipment for worker protection. Based on environmental monitoring data we have found that worker exposures are very limited at this site and that the main danger to the cleanup worker is heat stress resulting from the use of protective equipment. The potential value of using biomarkers under these conditions is also examined.

Challenges for Worker Monitoring at the Hanford Site

The Hanford Site, a former nuclear processing site, is located in a remote area in the State of Washington in the United States. The Hanford Site is extensive, covering 564 square miles, and is bounded by the Columbia River and the city of Richland. The waste tank disposal area is of particular concern and consists of 177 large tanks containing between 500,000 and 1,000,000 gallons each. One hundred forty-nine of these tanks are single-shelled tanks, of which 67 are leaking or are assumed to have leaked because of monitoring data. The site also has 28 double-shelled tanks, none of which are assumed to have leaked.

This waste site presents unique problems for worker monitoring.

Multiple chemicals are present, and new chemicals are being formed by the mixtures of chemicals present. There is ongoing exothermia in the tanks as a result of both chemical reactions and radiation events. A very ambitious and intense scientific effort is currently under way to characterize this site. To date, approximately 10,000 industrial hygiene samples have been taken by the Hanford Environmental Health Foundation/Westinghouse Hanford Company via source, area, and personal dosimetry monitoring. The available source data do not indicate a high level of environmental contamination.

Monitoring is also accomplished by the Head Space Monitoring Toxic Evaluation Group, which consists of a large group of scientists including chemists, genotoxicity experts, geneticists, toxic health risk evaluators, and physicians. Its mission is to evaluate sampling data on the head space of the tanks. The headspace is sampled at three levels to avoid the influence of high concentrations of certain chemicals that occur at only one level. Further objectives of this group include evaporating the liquid content of the tanks, using a process of vitrification to enclose toxic substances and building six new double-shelled tanks.

During this process, 2,800 of the 17,000 workers at Hanford are classified as "hazardous material workers," which means that they enter the tank area. It is not expected that they will be subjected to similar exposures. For example, there is a potential explosion problem due to the presence of flammable gas in about 25 tanks. The possibility of explosions due to ferrocyanide is of concern for 20 tanks. There are 9 tanks in which there is an ongoing exothermic reaction such as between a nitrate and organic chemicals. Tank C106 currently has the highest heat load. All tanks are scheduled to be sampled and characterized within three years.

Currently, Westinghouse maintains a policy of using conservative control measures to prevent worker over-exposure to vapors during routine entry into areas with single-shelled tanks. These measures have consisted of conducting area sampling or using supplied air for entry into areas judged to pose a moderate potential for worker over-exposure. There is also a policy of mandatory use of supplied air during entry into tank areas judged to pose the highest potential for worker over-exposure. At the moment Westinghouse is attempting to remove some of the workers in the low hazard areas from respirator protection and supplied air. It has become apparent that the danger to these workers is greater from heat and orthopedic and ergonomic factors associated with the use of personal protective equipment than from their exposures.

However, until the headspace of all of the tanks is characterized, a sense of total hazard characterization will not exist. New chemicals, e.g., nitriles, will continue to evolve because of reactions among the mixtures of the existing chemicals. Even though characterization may be complete today in one given area, this characterization may change in coming years. How can we monitor for potential carcinogens and changing chemicals that may not be carcinogens? Should we monitor all of the chemicals in the tank headspaces or monitor only selected chemicals, even though the toxicity of all of those selected chemicals may not be known?

Monitoring Hazardous Waste Workers

In addition to the heat stress associated with the protective equipment, other health risks that are encountered by hazardous waste site workers involve exposure to both chemicals or radiation.

The possible benefits of monitoring such workers with biomarkers include:

- If the exposure assessment raises concern for the potential for overexposure, biologic monitoring may provide additional information;

- If exposure is via inhalation only, personal breathing zone monitoring would be sufficient to assess exposure. However, if dermal exposure is also of concern, air monitoring is not an adequate measure of the total workplace exposure;

- Environmental monitoring does not adequately assess exposure. This may be the case particularly when respirators or other personal protection equipment are in use (as is the case at this site), when absorption is uncertain, when individual variability is very wide, or when exposure fluctuates rapidly over time;

- When environmental monitoring is not feasible, which may occur when work is done in enclosed areas or when protective clothing is used, biological monitoring may be useful;

- Biological monitoring may be preferred when cumulative toxins are present, especially those with long half-lives (these may include metals that can bind to tissue and organic chemicals that are highly fat-soluble and poorly metabolized, such as dioxins).

- Biomonitoring of total exposure is clearly the desirable method in certain cases such as cholinesterase monitoring for exposure to many organophosphate pesticides, methemoglobin measurements as an indicator of exposure to multiple inducers, and levels of DNA adducts as an indicator of exposure to multiple neoplastic drugs.

Medical surveillance of the hazardous waste worker is a special challenge. Because of potential exposure to hundreds of substances, usually at low levels, and to mixtures of unknown composition, this group is clearly different from workers in a factory. In part, the challenge lies in the fact that conventional medical surveillance relies on standard medical examinations and laboratory tests that perform best in the evaluation of sick individuals. These procedures are simply not sensitive enough to detect the early abnormalities in groups of people who are basically well.

The protection of hazardous waste workers, whether indoors or out, has come to rely heavily on the use of personal protection equipment. Usually there is little consideration given to ergonomic factors, heat stress, isolation, etc., which are additional stresses encountered by workers using this kind of equipment. The Hanford Site has gone through an elaborate year-long evaluation of environmental, source, personal, and, headspace monitoring data to define the relative risk in order to remove certain workers from the use of personal protective equipment (PPE).

One of the concerns with the use of PPE is heat stress. There may be individuals who are vulnerable to heat stress and for whom heat stress embodies significant cardiovascular risk. If so, how can they be identified and evaluated? Does the use of PPE increase the susceptibility of these individuals to heat stress? Have these issues been addressed as rigorously as necessary? Sometimes the duration of usage (the "stay time") for an individual in a double PPE is shorter than the time it takes to put the equipment on.

The Medical Surveillance Program at Hanford

The medical surveillance program at Hanford was developed around the concept of homogeneous exposure groups (HEGs). HEGs are groups of employees with similar job functions and similar, although not necessarily identical, exposures. Members of HEGs are surmised to share risk profiles. Therefore, they should perhaps also be considered as homogeneous risk groups. How is risk perceived for a HEG? The background data that are necessary to form such a group consist of job descriptions, job task analyses, available industrial hygiene data, an updated and useful occupational health history and the selection of appropriate monitors in medical surveillance to be able to evaluate change as it occurs over a period of time.

The first group that was selected for medical surveillance as a homogenous exposure group at Hanford was journeymen painters. The job descriptions, job task analysis, and qualifiable data from industrial hygiene were available to identify the populations and probable exposures. However, no quantifiable data were yet available. Two hundred and eighty people were in this program, and a medical surveillance profile and data monitors were developed to follow this group over time.

In the tank farm group, there are 2,800 people. They are not all hazardous waste cleanup workers, but they all must have a hazardous material examination because of ingress and egress into the tank farm. This group includes scientists, clerical workers, some nuclear process operators, radiation protection technicians, and the multiple crafts persons who are found in such an environment. Initially, three groups were identified on the basis of risk: 1) Nuclear process operators and radiation protection technicians, 2) maintenance crafts persons, and 3) clerical staff and scientists. For each of these categories, there are multiple job descriptions, multiple job task analyses and multiple, although different, exposures. However, the qualification of risk was the criterion for the division of the three groups. Medical surveillance was developed on the basis of industrial hygiene and safety data and from prior medical surveillance exams which qualified all of these people as hazardous waste workers. All stressors and risks were considered for the three groups: exposure to chemicals (known and unknown), heat or cold stress, ergonomic or orthopedic factors, psychological stress, etc.

An updated occupational health history was compiled for each worker using the NIOSH occupational health history format. Many of the workers had transferred jobs over their 20-year period at the Hanford Site. People were generally placed into medical surveillance programs by their immediate supervisors in the contracting shop. Until recently, there was little ability to modify that process. For workers who changed jobs over a 20-year period, the horizontal medical history is often not well maintained. This has happened for example with prior beryllium and asbestos workers.

We are currently in the process of developing quantifiable baseline industrial hygiene data which will be shared by all of the industrial hygienists to refine the HEG rather than just place workers into the existing HEGs. Interpretation of qualitative and historical data and clinical judgment are difficult to determine objectively, but these form the necessary basis for risk estimates to place people in the medical surveillance program. In most cases, available industrial hygiene data are insufficient to make an adequate estimate of hazard-based risk for the work force.

CONCLUSIONS

Biomarkers may prove useful in refining the estimate of risk when validation, sensitivity, specificity, and positive predictive value are improved. They are needed because they identify specific points in time in the continuum from exposure to disease. They do not equate with disease but are meant to identify people at risk of disease before that disease is present. This concept runs counter to the previous common thinking in the medical community which was that a person either had a disease or did not. Biomarkers represent a truly preventive measure in the specialty of occupational preventive medicine. We all look forward to the time when biomarkers are used commonly and selectively for whatever stressor is of concern to the individual. Hanford is moving into the use of targeted, specific examinations, as mandated by the Americans with Disabilities Act (ADA), to identify and survey people for the actual things to which they are exposed in the workplace. If we accomplish this goal, biomarkers as a specific, targeted, reliable, and reproducible tool will be a necessary arm for a complete occupational medicine program.

RECOMMENDED LITERATURE

Aitio, A. 1992. Biological monitoring at the Institute of Occupational Health, Helsinki, Finland. Scandinavian Journal of Workers Environmental Health 18:69-71.

Allessio, L. 1992. Reference values for the study of low doses of toxic metals, Occupational Health Institute University of Brescia, Italy. Scientific Total Environment 9:120.

Chinski, A. 1990. Occupational medicine-state of the art reviews. Medical surveillance in the workplace-legal issues. July-September 1990, Volume 5, Number 3, p. 457.

Department of Labor, Occupational Safety and Health Administration, 29 CFR Part 1910.120 Hazardous Waste Operations and Emergency Response, final rule, paragraph F. Medical Surveillance.

Gochfeld, M. 1990. Occupational medicine-state of the art reviews. Hazardous waste workers. January-March 1990, Volume 5.

Harrington J. M., and F. S. Gill. 1987. Occupational Health, Blackwell Scientific Publications, Second Edition.

Ladou, J. 1990. Occupational medicine-state of the art reviews. Rhythmic variations in medical monitoring tests. July-September 1990, Volume 5, Number 3, p. 479.

Monster, A. C., et al. 1991. Biological exposure and/or effects, limits, facts, fallacies and uncertainties: practical aspects from the Journal of the Society of Occupational Medicine, Coronel Laboratory, Amsterdam, The Netherlands, Journal Society of Occupational Medicine 41:60-63.

Rempel, D. 1990. Occupational medicine-state of the art reviews. Medical surveillance in the workplace. July-September 1990, Volume 5, Number 3, p. 435.

Rom, W. N. 1992. Environmental and Occupational Medicine, Second Edition, Little Brown & Company.

Rosenberg J., and D. Rempel. 1990. Occupational medicine-state of the art reviews. Biological Monitoring. July-September 1990, Volume 5, Number 3, p. 491.

Rosenstock and Cullins. 1986. Clinical Occupational Medicine, Saunders Blue Book Series.

Schulte, P. A. 1991. Contribution of biological markers to occupational health. American Journal of Industrial Medicine 20:435-446.

Sugita, M. 1991. Quality control program on biological monitoring by Japan federation of occupational health organizations. Internal Archives of Occupational and Environmental Health 62:569-577.

Biomarkers and Occupational Health: Progress and Perspectives. 1995. Pp. 140-147

Integrating Biomarkers into Health and Safety Programs

Morton Corn

The spectacular progress in biomarker research during the past 5-10 years has been associated mainly with utilization of these tools in epidemiology, as summarized in the volume *Molecular Epidemiology: Principles and Practices* (Schulte and Perera, 1993). Extension beyond epidemiology into considerations of diagnostic medicine were explored in *Biological Markers in Immunotoxicology*, a publication of the National Research Council (1992). However, it is important to note the various concerns of occupational health practitioners concerning the utilization of these tools in organization health and safety programs. Specifically, members of the traditional "team of occupational health practice," which includes the occupational health physician, occupational health nurse, industrial hygienist, and safety professional are concerned with the implications of these advances for day-to-day programs in the delivery of occupational health services. This paper will address the nature of the occupational health and safety program and the program elements where biomarkers may have possible utilization. The paper will also outline some criteria that these tools must meet if they are to be appropriately utilized.

Biomarkers generally include biochemical, molecular, genetic, immunologic, or physiological signals of events in biological systems (NRC, 1992). This discussion will not differentiate among the specific biomarkers, but will generally use the term biomarker to denote any of these categories.

The Importance of a Health and Safety Program

Any occupational setting should develop a health and safety program (HASP) designed to minimize risk in the occupational environment. Its purpose is prevention. The HASP is adapted to the individual risks of the organization adopting the program. A list of typical program elements in an occupational health program is shown in Table 1. The profile of program elements derives primarily from the second entry in Table 1, namely the job/task hazard analysis. This is a crucial first step in determining the risks of the particular setting.

TABLE 1. SELECTED HEALTH AND SAFETY PROGRAM ELEMENTS

o	POLICY STATEMENT
o	JOB/TASK HAZARD ANALYSIS
o	EMPLOYEE EDUCATION AND TRAINING
o	HEARING CONSERVATION PROGRAM
o	EXPOSURE ASSESSMENT
o	CHEMICAL CONTROL PROCEDURES
o	RADIATION SAFETY PROGRAM
o	FIRE PREVENTION PROGRAM
o	LABORATORY SAFETY PROGRAM
o	MEDICAL PROGRAM
o	CONTROL PROGRAM
o	EMERGENCY RESPONSE
o	ERGONOMICS

In addition to direct observation of the job/task, there are other sources of insight into the potential risks of different jobs/tasks in an organization. Table 2 presents sources of risk data that may be available in the occupational setting (Corn, 1991). Lost-time records and frequency and severity rates are particularly helpful for assessing traumatic injury, while workmen's compensation records can be useful for identifying occupational diseases. Various "systems analysis" approaches to estimating risk on the job are also available. These include the "what if" method, failure-mode analysis, fault-tree analysis, similarity analysis, and hazardous-operations analysis. These approaches are often referred to as the techniques of hazard analysis and are most frequently used with low probability, high consequence events.

The use of job/task analysis involves the physician, the safety professional and/or industrial hygienist observing the employee at work to determine the potential hazards of their activity. Prior to this observation, the chemical agents to be utilized, the processing activities that will be involved, etc., are examined. A list of potential hazards/risks is then formulated. This is done for each job title or for individual employees, if job titles encompass highly divergent tasks. Job/task analysis differs from exposure assessment, as used here, in that exposure assessment has a quantification function. For example, exposure assessment for a chemical would involve estimating the exposure concentrations, predicting the routes of entry into the body and the time of exposure(s) and estimating the dose to the body. Exposure assessment is differentiated from exposure monitoring in that the latter is usually routine and repetitive to ensure that the agent control systems are effective. Exposure assessment can be viewed as a validation of the job/task analysis.

TABLE 2. SOURCES OF RISK ESTIMATES

1. LOST-TIME RECORDS (INJURY AND ILLNESS)
2. FREQUENCY AND SEVERITY RATES
3. REPORTED "NEAR MISSES"
4. WORKMEN'S COMPENSATION RECORDS
5. PROPERTY-LOSS RECORDS (FIRE, FLOOD, AND OTHER ACTS OF NATURE)
6. ONGOING SURVEILLANCE PROGRAM:

    o   MORTALITY
    o   MORBIDITY
    o   BIOMARKERS:

    BLOOD LEAD
    URINE MERCURY
    HEARING LOSS
    OTHERS?

7. PERCEPTION OF EMPLOYEES

The category of Chemical Controls includes many components, as shown in Table 3. These controls encompass a "cradle-to-grave" concept analogous to the handling of hazardous wastes and include exposure assessment, which is the first potential application of biomarker utilization in the HASP. A biologic marker of chemical sensitivity, in general, or to a specific chemical species, presents the promise of identifying those persons at increased risk of developing certain diseases or of exhibiting chemical sensitivity (Committee on Biological Markers of the National Research Council, 1987). The exposure assessment also offers a potential opportunity for the utilization of biomarkers that confirm the presence of disease. A disease marker is a measurable indicator of a biological or biochemical event that either represents a subclinical stage of disease or is a manifestation of the disease itself (Wogan, 1989). It would be most useful to have these markers as indicators of the pre-disease state, namely as a screen for preclinical disease. However, progress in this area has been limited because of the non-specificity of many markers in this category (Hulka et al., 1990).

TABLE 3. CHEMICAL CONTROLS

o   MATERIAL SAFETY DATA SHEETS
o   HAZARD COMMUNICATION
o   ORDERING RESTRICTIONS
o   RECEIPT, STORAGE, AND LABELLING
o   APPROPRIATE AREA OF USAGE
o   EXPOSURE ASSESSMENT/MONITORING
o   PROPER USAGE/CONTROLS
o   WASTE DISPOSAL
o   RECORDKEEPING

Another category of markers for occupational health practice is markers of exposure, which ideally can indicate the dose to the body or target organ more accurately than exposure monitoring methods external to the body. Exposure monitoring has multiple purposes including determination of the equipment effectiveness and the spatial/temporal variations of agents in the work environment; these are goals that would not be fulfilled by the biomarkers envisioned for measures of exposure, dose, or disease in the worker. Exposure monitoring and biomarker use should be viewed as complementing each other in practice, rather than the marker replacing exposure monitoring. The goal of exposure monitoring is prevention; the positive response in biomarker screening indicates that prevention has failed and exposure has occurred. A positive result for a biomarker of exposure should trigger re-examination of HASP control interventions.

The Medical Program in the HASP is the program element with potentially the greatest involvement in application of biomarkers. The occupational health physician would be the safety professional involved in receipt of information pertaining to biomarker confirmation of susceptibility, exposure, or disease.

Biological Monitoring

In contrast to the discussion on biomarkers, biological monitoring is defined as a systematic, repetitive health-related activity designed to lead, if necessary, to corrective action. A further explanation is that biological monitoring is the measurement or assessment of agents that are present in blood or other tissues, or are secreted, excreted, or exhaled in expired air (or any combination of these), in order to evaluate the exposure in terms of health risk, compared to an appropriate reference period. The field of occupational health has experience with a host of such agents that can be measured. Table 4 presents a list of selected current chemicals from the thirty listed by the Committee on Threshold Limit Values of the American Conference of Governmental Industrial Hygienists (1993) as chemicals which have biologic exposure indices. These indices would not in all cases meet the definition of biomarkers as either agents of dose, response, or susceptibility. However, they do confirm exposure, and there are guidelines believed to be valid for preventing disease when the agent reaches a given concentration in the appropriate biologic media. However, these are not all equally valid indices and the interpretation of the results is not always simple. There are enormous biological variations between individual workers. The point is that they have proven to be useful in a practical health and safety program.

It should be noted that biological monitoring with invasive techniques is less desirable than, for example, monitoring of exhaled breath for alcohol content. The latter process is more acceptable to employees than taking a blood sample for lead, for example.

Individual biological monitoring for a given substance can be quite complex. As an example, both blood and urine can be monitored for mercury. The urine content of mercury reflects the body burden being processed by the kidney and the blood level reflects recent exposure. Since the earliest sign of mercury damage is to the central nervous system (CNS), urinary monitoring is really not a good early indicator of that potential. The quality of the correlation of urine and blood mercury levels is a matter of current controversy. Therefore, a mercury biological monitoring program utilizing the urine may not avoid CNS effects brought about by short-term higher level exposures. Despite the nuances in interpretation of these biological monitors, which is both a science and an art for physicians and toxicologists, it is acknowledged that biomarkers are another potentially useful tool in a progression of tools available for monitoring the work population.

TABLE 4. SELECTED BIOLOGICAL EXPOSURE INDICES (BEIs) FROM THE
AMERICAN CONFERENCE OF GOVERNMENTAL INDUSTRIAL HYGIENISTS
GUIDELINES, 1993-1994

DETERMINANT	BEI	NOTATION
CADMIUM		
URINE	5 µg/g CREATININE	B
BLOOD	5 µg/L	B
BENZENE		
TOTAL PHENOL IN URINE	50 mg/g CREATININE	B, NS
LEAD		
BLOOD	50 µg/100 ml	B
URINE	150 µg/g CREATININE	
MERCURY		
TOTAL INORGANIC		
MERCURY IN URINE	35 µg/g CREATININE PRESHIFT	B
TOTAL INORGANIC		
MERCURY IN BLOOD	END OF SHIFT AT	
	END OF WORKWEEK	B
	15 µg/L 35	B
NITROBENZENE	5 mg/g CREATININE	NS

B=BACKGROUND
NS=NON-SPECIFIC INDICATOR

Criteria for Biological Markers Having Potential Utilization in HASPs

Table 5 presents minimum criteria for the use of a biomarker in a HASP. The first criterion is a valid and preferably non-invasive method for sampling and analysis, as noted previously. Reproducibility refers to the ability to achieve similar or comparable results within the same arena of investigation and, ideally, by other investigators using similar methods.

The criteria of sensitivity and specificity would eliminate Type I and Type II errors, namely a false negative result when the characteristic is present or a false positive result when the characteristic is not present, respectively. In matters involving the potential impact of a biomarker for susceptibility, Type I and II error rates of less than 1/1000 are not unreasonable expectations by potential subjects. Biological fluid monitoring, as currently used in the private sector, cannot meet this criterion.

TABLE 5. MINIMUM CRITERIA FOR ACCEPTABLE USE OF
BIOMARKERS IN THE WORKPLACE

---

1. VALID AND PRACTICAL (PREFERABLY NON-INVASIVE) METHODS OF SAMPLING
    AND ANALYSIS

2. REPRODUCIBILITY

3. SENSITIVITY AND SPECIFICITY

4. CLEAR AND UNEQUIVOCAL INTERPRETATION OF THE MEASUREMENT IN
    TERMS OF THE SUSCEPTIBILITY, BIOLOGICALLY ACCEPTABLE DOSE, BIOLOGIC
    RESPONSE, OR PRECLINICAL/CLINICAL DISEASE

5. WORKER INFORMED CONSENT ACHIEVABLE

6. DEFENSIBLE ETHICAL, LEGAL PROCEDURES FOR UTILIZATION

The fourth criterion is that interpretation of the results must be clear and unequivocal. This criterion is intimately associated with the preceding three criteria because it depends so strongly on them.

Criteria 5 and 6 are procedural criteria and recognize the need for socially acceptable methods to implement the utilization of biomarkers meeting the previous four criteria. It is conceivable that the scientific, technical criteria could be met, but that consensus could not be reached on defensible procedural criteria. Although not specified in Table 5, informing the subject of results is inherent in the sixth criterion.

Only when the above criteria are met should biomarkers be considered for inclusion in HASPs.

CONCLUSIONS

Biomarkers offer substantial improvement as a component of HASPs. They can supplement current medical surveillance and exposure monitoring program elements, in particular, to prevent disease on the job. Specific criteria are proposed for biomarkers before they are considered for implementation in HASPs. The different categories of biomarkers will probably not meet with equivalent levels of difficulty and controversy in their utilization; biomarkers of susceptibility will probably be most controversial and difficult to integrate into ongoing HASPs.

REFERENCES

American Conference of Governmental Industrial Hygienists. 1993. 1993-1994 Threshold Limit Values for Chemical Substances and Physical Agents and Biological Exposure Indices. Cincinnati, Ohio.

Committee on Biological Markers of the National Research Council. 1987. Biological Markers in Environmental Health. Environmental Health Perspectives 74:3-9.

Corn, M. 1991. Risk Awareness: Approaches to Selecting Employee Risks for Reduction. Pp. 33-37. Proc. 12th NIH Research Safety Symposium. Washington, D.C. March 22-23, 1990. NIH Pub. No. 91-3200. Department HHS, PHS, NIH.

Hulka, B. S., T. Wilcosky, and J. D. Griffith. 1990. Biologic Markers in Epidemiology. Oxford University Press, New York.

National Research Council, National Academy of Sciences. 1992. Biologic Markers in Immunotoxicology, Washington, D.C.

Schulte, P. A., and F. P. Perera. 1993. Molecular Epidemiology: Principles and Practices. Academic Press, New York.

Wogan, G. N. 1989. Markers of Exposures to Carcinogens. Environmental Health Perspectives 81:9-17.

Biomarkers and Occupational Health: Progress and Perspectives. 1995. Pp. 148-160

Clinical Applications of Biomarkers in Occupational Medicine

Robert J. McCunney

The diagnosis of occupational/environmental illness can be imprecise and challenging since many illnesses have similar clinical patterns. For example, the oat cell carcinoma of the lung that has been reported to be associated with exposure to bis-chloromethyl ether is similar in both treatment and prognosis to the same histological tumor that develops in cigarette smokers. Asthma associated with nonoccupational substances such as dusts and mites resembles asthma due to exposure to isocyanates. In addition, the dementia that has been described in some solvent-exposed workers resembles the presenile dementia of Alzheimer's disease.

In the diagnosis of an occupational illness, tremendous emphasis is placed on the person's occupational history, the quality of which may vary considerably, depending upon numerous factors, such as the availability of good exposure data and the historical recollection of the patient (Goldman and Peters, 1981). Often, there may be a conflict in the information that is presented in the examination room versus that described by management or observed at the worksite by the physician. As a result of challenges in formulating a definitive diagnosis of an occupational illness, there is active interest in the role of biomarkers on the part of federal agencies, labor, management, and the legal system. Recently, claims have been made for biological monitoring in the context of a variety of toxic torts.

There are several reasons why such a great deal of emphasis is currently placed on biomarkers. Biomarkers may enhance the diagnostic accuracy of occupational and environmental illnesses and ultimately result in the prevention of disease in other individuals. Compensation for persons adversely affected by these exposures may be more equitable. In addition, biomarkers are likely to enhance the understanding of the dose-response relationship between exposure to a hazard and an illness. Ultimately, their use may help to evaluate the effectiveness of various control measures.

148

This paper will focus on the clinical utility of biomarkers from the perspective of an occupational physician who has served primarily in hospital settings.

TYPES OF BIOMARKERS AND CLINICAL CHALLENGES

Although the terms biomarkers, biomonitoring, medical monitoring, and medical surveillance are often used interchangeably, biomarkers are used here to refer to clinical tests that are specific to an occupational or environmental exposure or illness. Biomarkers were defined by the National Research Council in 1982 as follows:

1) Biological markers of *exposure* reflect an exogenous substance or its metabolite that is measured in a compartment of the body.

2) Biological markers of *effect* refers to measurable alterations that can be recognized as an established or potential health impairment.

3) Biological markers of *susceptibility* are indicators of a person's ability to respond to a xenobiotic challenge.

Biological Markers of Susceptibility

Biological markers of susceptibility are of great interest, because of their potential to indicate a substantial risk of illness as a result of an exposure to a hazard. These markers are primarily directed toward the risk of malignancy. For the most part, however, aside from research settings, these tests have not had clinical practicality. In fact, the National Advisory Council for Human Genome Research (1994) recently stated that it is premature to offer DNA testing or screening for cancer predisposition outside of a carefully monitored research environment.

Biological Markers of Effect

Biological markers of effect have been used in occupational medicine for many years. These types of markers refer to clinical chemistry tests that are often used in non-occupational settings. Measurement of liver function enzymes is a notable example. Many different agents used in industry and in various occupations have been associated with occupational hepatitis (McCunney and Harzbecker, 1992). Hepatitis can be caused by viruses, alcohol, medication, and drugs as well as by exposure to occupational agents. The advent of biomarkers, however, may shed light on the causes of various liver abnormalities. Unfortunately, even in the diagnosis of hepatic cirrhosis, despite the fact that there are biochemical, histological, imaging, physiologic, and physical abnormalities, there is no specific pattern associated with this illness that can be clearly associated with exposure to an occupational agent.

Physicians in the clinical practice of occupational/environmental medicine are occasionally asked for an opinion regarding the value of a biological marker, either in evaluating an illness or an exposure to a particular hazard.

Clinical challenges facing the physician in evaluating potential occupational and environmental illness fall under the following categories:

- Acute illness or symptom(s)
- Subacute illness
- Non-specific illness
- End organ damage
- Risk of future illness

Acute occupational illness or symptoms may be due to exposure to solvents, such as trichlorethylene, perchloroethylene, 1,1,1-trichloroethane as well as to styrene and other substances. Biomarkers can be helpful in evaluating *subacute illness*, such as asthma, especially in the context of exposure to isocyanates, platinum salts, nickel compounds and epoxy resins. *Non-specific illnesses*, such as those described as multiple chemical sensitivity, an as yet unvalidated diagnosis, may also call into question the value of specific markers. Symptoms associated with a hazardous waste site or pesticide exposure are other examples that challenge the clinical acumen of an occupational physician.

In the context of *end organ damage*, such as hepatic cirrhosis, peripheral neuropathy, various lung diseases, or leukemia, the physician may order specific tests that might help clarify the cause of the ailment. *Risk of future illness*, secondary to exposure to asbestos or pesticides in the home, may challenge the physician's expertise and raise the question of the clinical utility of biomarkers.

Biological Markers of Exposure

In the occupational setting, situations in which biomarkers are probably most useful are cases of exposure to metal dusts and solvents and occasionally pesticides. Three exposure scenarios, acute, chronic, and retrospective, often prompt requests for biomarker testing. In the case of *acute exposures* which result from a laboratory spill or a malfunctioning hood that releases a toxic substance such as cadmium dust, biomarkers can be extremely helpful in assessing the level and degree of exposure to the hazard. Biomarkers can also be effective in cases of exposure to barium, chloroform, or gallium arsenide among others. Transformer spills and leaks of polychlorinated biphenyls (PCBs) or releases of ethylene oxide from an improperly operated sterilizer are other situations in which biomarker testing would be useful.

Chronic exposure may result from hazards present in the water supply. Benzene, lead, trichloroethylene, and even molybdenum have been contaminants of water. In the workplace, exposure to asbestos, polycyclic aromatic hydrocarbons from carbon black, or cobalt in metal refining operations, often challenge the physician for specific recommendations. Other examples of chronic exposure include arsenic in hazardous waste workers and molybdenum exposure from producing catalysts for the chemical industry.

Retrospective exposure assessments are difficult to conduct, not only in epidemiological studies, but also in the clinical setting. People may be concerned about previous exposure to lead, beryllium, asbestos, and other hazardous materials.

Screening/Biological Monitoring

In the practice of occupational medicine, biological monitoring may be used as part of a medical surveillance program. Some standards promulgated by the Occupational Safety and Health Administration, such as those for arsenic, cadmium and lead, require medical surveillance. Medical surveillance refers to the systematic collection, analysis and dissemination of health-related information on groups of people in order to identify occupational disease (Harber et al., 1994). The effectiveness of medical surveillance is dependant on the medical screening program, which attempts to identify a medical condition at a stage that is early enough that intervention can slow, halt, or reverse the condition.

Screening techniques employ questionnaires, examinations, and laboratory testing. Screening criteria focus on the population at risk and attempt to identify the disease in its latent phase, **not** when symptoms have already begun to appear. Adequate follow-up should be ensured; the screening test should be both valid and reliable.

A variety of medical surveillance requirements are included in some of the OSHA standards. OSHA standards in place in 1994 include 17 with medical provisions, but only three (those for arsenic, lead, and cadmium) require specific medical monitoring. The medical surveillance requirements for lead are very specific.

Biological monitoring is similar to screening; that is, when environmental exposure occurs, some biological parameter is examined ideally before adverse effects appear.

In the context of using biological markers for assessment of either illness or exposure, a number of key factors should be noted. Biological monitoring attempts to determine the internal dose of a substance. Usually, either the agent itself or one of its metabolites is measured in the blood, urine, or exhaled air. In contrast to routine clinical chemistry tests that are well automated and standardized, the analytical techniques of laboratories that perform biological monitoring vary considerably. In fact, biological monitoring procedures are not well standardized, in comparison to routine clinical testing protocols.

In determining the need for biological monitoring in any occupational or environmental setting, a number of factors should be addressed. For example, there should be some hypothesis developed for assessing the potential exposure/disease relationship. Decisions need to be made as to what group(s) or people will be tested, what parameter (the agent itself or one of its metabolites) will be assessed, what type of biological specimen (blood, urine, etc.) will be analyzed, and the timing of the sampling. Other factors include the availability of appropriate reference limits for the substance as well as its metabolic profile. In addition, there must be some consideration of the potential impact of dietary deficiencies that can alter the absorption of some hazardous substances. For example, in individuals with either calcium or iron deficiency syndromes, the absorption of cadmium and lead may be enhanced.

In evaluating the results of biological monitoring, a number of references are available (Baselt, 1988; Lowry, 1986). In addition, in 1982, the American Conference of Governmental Industrial Hygienists first proposed their biological exposure indices (BEIs) (McCunney, 1994) (See Table 1 and chapter by V. Thomas). These criteria were compiled by an expert committee based on a review of the medical literature to determine parameters appropriate for monitoring the hazard. These guidelines refer to levels of the agent itself or its metabolites in urine, blood, or exhaled air that are associated with exposures to each substance at

TABLE 1. AIRBORNE CHEMICALS FOR WHICH BIOLOGICAL
EXPOSURE INDICES (BEI) HAVE BEEN ADOPTED

Airborne chemical	Determinant
Aniline	Total p-aminophenol in urine Methemoglobin in blood
Benzene	Total phenol in urine Benzene in exhaled air
Cadmium	Cadmium in urine Cadmium in blood
Carbon disulfide	2-Thiothiazolidine-4- carboxylic acid in urine CO in end-exhaled air
Chlorobenzene	Total 4-chlorcatechol in urine Total p-chlorophenol in urine
Chromium (VI)	Total chromium in urine
N.N-Dimethylformamide	N-Methylformamide in urine
Ethylbenzene	Mandelic acid in urine Ethylbenzene in end-exhaled air
Fluorides	Fluorides in urine
Furfural	Total furoic acid in urine
n-Hexane	2,5-Hexanedione in urine n-Hexane in end-exhaled air
Lead	Lead in blood Lead in urine Zinc protoporphyrin in blood
Methanol	Methanol in urine Formic acid in urine
Methemoglobin inducers	Methemoglobin in blood
Methyl chloroform	Methyl chloroform in end-exhaled air Trichloroacetic acid in urine Total trichloroethanol in urine Total trichloroethanol in blood

TABLE 1. AIRBORNE CHEMICALS FOR WHICH BIOLOGICAL
EXPOSURE INDICES (BEI) HAVE BEEN ADOPTED (CONT.)

Methyl ethyl ketone	MEK in urine
Nitrobenzene	Total p-nitrophenol in urine Methemoglobin in blood
Organophosphorus cholinesterase inhibitor	Cholinesterase activity in red cells
Parathion	Total p-nitrophenol in urine Cholinesterase activity in red cells
Pentachlorophenol (PCP)	Total PCP in urine Free PCP in plasma
Phenol	Total phenol in urine
Styrene	Mandelic acid in urine Phenylglyoxylic acid in urine Styrene in venous blood
Toluene	Hippuric acid in urine Toluene in venous blood Toluene in end-exhaled air
Trichloroethylene	Trichloroacetic acid in urine Trichloroacetic acid and trichloroethanol in urine Free Trichloroethanol in blood Trichloroethylene in end-exhaled air
Xylenes	Methylhippuric acids in urine

*   Adapted from McCunney J. M. 1994. Industrial Hygiene. Pp. 321-333 in: *A Practical Approach to Occupational & Environmental Medicine*. McCunney R. J., ed. Little Brown, Boston, 1994.

the threshold limit value (TLV). In some cases, the concentrations noted in these indices are reflective of exposure and not adverse health effects.

The Use of Biomarkers in Evaluating Illness

Biological markers can be extremely helpful in evaluating acute symptoms related to exposure to some solvents and metals. For example, for a laboratory worker with symptoms resulting from styrene released from an improperly operating hood, the level of exposure can be assessed by measurement of mandelic acid in the urine. Clearly, acute symptoms should be treated and the source of exposure should be repaired or eliminated; however, measuring the amount of material that is absorbed can provide valuable prognostic information to the attending physician.

Measuring the levels of one of the metabolites of trichloroethylene in the blood or urine of a factory worker who dipped his hands in a degreasing fluid is another example of an effective use of biomarkers (McCunney, 1988). Similarly, performing a blood lead test in a person who conducted renovation work that included exposure to lead-based paint, can be extremely helpful in evaluating symptoms and proposing treatment.

Subacute illnesses, such as asthma, can also have the diagnostic accuracy enhanced through the use of some markers such as radioallergosorbent testing (RAST) (Accetta and Farnham, 1994). RAST testing can be performed for a variety of substances including molds, isocyanates, and latex as well as mouse and rabbit epithelium. Skin tests to these agents can supplement the diagnostic accuracy of the RAST testing. An accurate diagnosis of asthma can facilitate proper advice to the patient and ultimately prevent future illness.

Some chronic illnesses can be more precisely diagnosed through biomarkers. For example, chronic beryllium disease can be differentiated from sarcoidosis with the use of a lymphoblast transformation test (Rossman, 1988, see also chapters by Saltini and by Rossman). Similarly, encephalopathy due to lead exposure can be differentiated from that associated with Alzheimer's disease through x-ray florescence (Kosnett, 1994).

In contrast, biomarker testing is of no apparent use in the diagnosis of a variety of common occupational illnesses, such as the pneumoconioses. Hypersensitivity pneumonitis, which is sometimes a concern in cases of poor indoor air quality, has no specific biomarker. Serum precipitins are merely markers of exposure. Microbial sampling of the ventilation system may need to be supplemented with appropriate skin testing.

In evaluating pulmonary alveolar proteinosis, electron dispersive spectroscopy has been used to identify the presence of silica particles in the lung biopsy of a worker exposed to silica dust (McCunney and Godefroi, 1989). Although this type of testing would not lend itself to routine use, the example illustrates how specific tests can be used to diagnose an occupational illness. This would be of value not only for providing clinical advice as to whether the person should return to work that entails a similar type of exposure, but also in ensuring that compensation and rehabilitative benefits will be appropriate.

Another specific example in which biological monitoring proved to be of benefit was in determining the cause of acute central nervous symptoms in a metal worker who was using trichloroethylene in degreasing operations. Since air sampling indicated that concentrations of trichloroethylene were within normal limits, it appeared that the patient's symptoms were due to other causes. Because the symptoms persisted, however, biological monitoring of urinary trichloroacetic acid levels was conducted after a full day's work; this level was extremely high, but the level returned to normal after the person was removed from work (McCunney, 1988).

Use of Biomarkers in Evaluating Exposure

The occupational medicine physician may be called upon to evaluate the use of biomarkers in acute exposure situations, such as spills, releases, and accidents. In these cases, biomarkers may be useful in determining the extent of exposure to the hazard and thus to properly advise the patient. In other cases, biological monitoring may be used to help evaluate the effectiveness of preventive control measures, such as administrative controls, industrial hygiene measures, and personal protective equipment.

Biomarkers can be effective in evaluating acute exposures to certain metals and solvents. For example, exposure to lead in a firing range or as a result of performing renovation work can be adequately assessed. Cadmium exposure in soldering operations as well as arsenic use in the laboratory can also be determined through biomarkers. In monitoring arsenic levels attention must also be given to recent seafood ingestion. Similarly, when symptoms occur secondary to an exposure that seems well controlled, monitoring for chloroform, perchloroethylene, trichlorethylene, styrene, and MBOCA have all proven effective. Monitoring blood levels of polychlorinated biphenyls (PCBs) after a PCB release has also proven effective.

In evaluating chronic exposure to some hazards, such as lead, the monitoring of lead levels along with levels of zinc protoporphyrin, or the monitoring of cadmium with blood levels and Beta-1 microglobulin are also effective. There appears to be some promise in monitoring polycyclic aromatic hydrocarbon exposure with 1-hydroxypyrene levels (Gardiner et al., 1992). These results are so dramatically affected by diet and cigarette smoking, however, that their utility is questionable. Monitoring for levels of molybdenum, for formic acid as a result of formaldehyde exposure and for methemoglobin as a result of aromatic amine exposure have not proven effective in most instances.

In summary, in the contemporary clinical practice of occupational medicine, biomarkers are more valuable in evaluating exposure than disease, and tend to be limited to evaluating acute exposures or acute illnesses, with rare exceptions.

CONCLUSIONS

Biological monitoring is likely to be effective in occupational medical for substances with relatively short half-lives (8-48 hours), especially when the pharmacokinetics are understood, the analytical techniques are valid, and appropriate reference limits are available (Knelp and Crable, 1988). It is further helpful if the results are correlated with health effects and if there is a specific metabolite available associated with the hazard itself. Very often a metabolite or its agent may be affected by non-occupational sources. For example, in monitoring exposure to xylene, the determination of methyl-hippuric acid is specific for xylene since this substance is not normally present in urine. On the other hand, in monitoring for toluene exposure, the determination of hippuric acid may not be useful because this compound can also be present in the urine as a result of exposure to preservatives used in wheat products. Urinary arsenic levels are well known to be increased by recent seafood ingestion, and polycyclic aromatic hydrocarbon exposures can be affected by diet and cigarette smoking.

Biomarkers can also be misused as well as lead to misleading results.

For example, some general toxicology screens that are commercially available may mislead people into believing that they are suffering from toxic effects, simply because an agent or a metabolite is measured in their blood stream. In fact, some practitioners have misused such results to indicate the presence of multiple chemical sensitivities, a diagnosis of dubious credibility.

Practitioners must be realistic in advising patients, especially regarding the risk of cancer. Too often, there is a big leap from the laboratory bench to the bedside. This caution is noted in the National Advisory Council for Human Genome Research (1994) described earlier. On the other hand, biological markers offer more precise information in evaluating exposure/illness relationships, even though they can be complicated by concurrent and multiple exposures at low dose. In fact, low-dose exposures present a more vexing challenge. For example, benzene exposure has long been monitored with urinary phenol levels, but this technique has not proven effective in monitoring low levels of benzene exposure.

At times, there may be hesitancy to use biomarkers by both management and labor. Labor may be concerned about potentially adverse consequences on jobs and employment that may occur as a result of participating in monitoring programs. Similarly, management may be concerned about liability that may result if over-exposures are noted.

Biological markers remain limited in their effectiveness for fast-acting and surface-acting agents, such as oxides of nitrogen, sulfur dioxide, ammonia, and ozone. Past exposures are not reliably measured through biomarkers with rare exceptions, such as the use of x-ray fluorescence to monitor accumulation of lead in the long bones (Kosnett, 1994) and the lymphoblast transformation test in monitoring previous exposure to beryllium (Rossman, 1988). Chronic illnesses, such as asthma and interstitial lung disease, offer limited potential for the use of biomarkers. In fact, biomarkers remain most effective for evaluating **acute symptoms** or **acute exposures** to metals and solvents.

Clearly, there is a need for greater research in the clinical applications of biomarkers. A number of methods may be employed to bridge the gap between the lab bench and the bedside. For example, in the proposed Occupational Safety and Health Administration (OSHA) generic medical surveillance standard, provisions might be included for serum storage, such as that which is performed for work with recombinant DNA and certain animal handling operations. Registries of exposed people, such as those developed by the Agency for Toxic Substances and Disease Registry, can also be used provided that proper ethical considerations are addressed. Basic biomedical research should continue, such as that directed towards indoor air quality problems where chemical markers are being developed for various bacterial agents. Finally, the efforts of the Human Genome project may prove to be helpful in elucidating susceptibility differences among individuals.

In conclusion, there are often complicated sociopolitical demands that arise in the context of exposure to hazardous materials that may affect scientific policy. There is a clear need for the practitioner to be straightforward and realistic in communicating information to workers as well as to management and not to unreasonably raise expectations. It is clearly necessary to differentiate the role of biological monitoring in the research versus the clinical setting.

REFERENCES

Accetta, D., and J. E. Farnham. 1994. Allergy and Immunology. Pp. 279-292 in: A practical approach to occupational and environmental medicine. R. J. McCunney, ed. Little Brown, Boston.

Baselt, R. 1988. Biological monitoring methods for industrial chemicals, Biomedical Publications, Davis, CA, 2nd edition.

Gardiner, K., I. Haleka, A. Clavert, C. Rice, and J. M. Harrington. 1992. The suitability of the urinary metabolite 1-hydroxypyrene as an index of polynuclear aromatic hydrocarbon bioavailability from workers exposed to carbon black. Annals of Occupational Hygiene 36:681-688.

Goldman, R. H., and J. M. Peters. 1981. The occupational and environmental health history. Journal of the American Medical Association 246:2831.

Harber, P., R. J. McCunney, and I. Monosson. 1994. Medical Surveillance. Pp. 358-375 in: A practical approach to occupational and environmental medicine. McCunney R. J., ed. Little Brown, Boston.

Knelp, T. J., and J. V. Crable. 1988. Methods for biological monitoring, American Public Health Association, Washington DC.

Kosnett, M. J. 1994. Factors influencing bone-lead concentration in a suburban community assessed by non-invasive K X-ray fluorescence. Journal of the American Medical Association 271:197-203.

Lowry, L. K. 1986. Biological exposure index as a compliment to the TLV. Journal of Occupational Medicine 28:578.

McCunney, R. J. 1988. Diverse manifestations of trichloroethylene. British Journal of Industrial Medicine 45:122-126.

McCunney, R. J., and R. Godefroi. 1989. Pulmonary alveolar proteinosis: A case report. Journal of Occupational Medicine 31:233-237.

McCunney, R. J., and J. Harzbecker. 1992. The role of occupational medicine in general medical practice. A look at the journals. Journal of Occupational Medicine 34:279-286.

McCunney, J. M. 1994. Industrial Hygiene. Pp. 321-33 in: A practical approach to occupational and environmental medicine. R. J. McCunney, ed. Little Brown, Boston.

National Advisory Council for Human Genome Research. 1994. Statement on use of DNA testing for presymptomatic identification of cancer risk. Journal of the American Medical Association 271:785.

Rossman, M. 1988. Lymphocyte proliferation test in berylliosis. Annals of Internal Medicine 108:697-692.

RECENT TECHNICAL ADVANCES
IN BIOMARKERS RESEARCH

Biomarkers and Occupational Health: Progress and Perspectives. 1995. Pp. 163-173

Mutant p21 Protein as a Biomarker of Chemical Carcinogenesis in Humans

Paul W. Brandt-Rauf, Marie-Jeanne Marion, and Immaculata DeVivo

Vinyl chloride (VC) is a known animal and human carcinogen which has been linked to the development of an unusual sentinel neoplasm, angiosarcoma of the liver (ASL) (ATSDR, 1993). In recent years, considerable insight has been gained into the potential carcinogenic mechanism of VC (Figure 1). VC is primarily metabolized in the liver by the cytochrome P450 2E1 system, and the electrophilic metabolites, chloroethylene oxide (CEO) and chloroacetaldehyde (CAA), are believed to be the most important in terms of carcinogenesis (ATSDR, 1993). CEO and CAA react with DNA bases to form adducts that are mutagenic in bacterial systems and mammalian cells (McCann et al., 1975; Huberman et al., 1975; Singer et al., 1988). Four DNA adducts have been identified in reactions with nucleic acids *in vitro* and in animals exposed to VC: 7-(2-oxoethyl) guanine; 1,N^6-ethenoadenine; 3,N^4-ethenocytosine; and N^2,3-ethenoguanine (ATSDR, 1993).

In animal experiments, the oxoethyl adduct is the major liver DNA adduct formed, representing 98% of all adducts, but it is the least persistent with a half-life of about 62 hours (Barbin et al., 1985). By contrast, the less common etheno adducts are highly persistent with a half-life of more than 30 days, suggesting that they are poorly recognized by the liver DNA repair system (Swenberg et al., 1992). Etheno adducts are known to be capable of causing miscoding (Barbin and Bartsch, 1986), and, in *in vitro* assays, the most frequent mutation induced by these adducts is a $G.C \rightarrow A.T$ transition which has been attributed specifically to the occurrence of the etheno guanine adduct (Cheng et al., 1991). One site where such $G \rightarrow A$ transitions are apparently produced is at the second base of codon 13 of *ras* oncogenes (Boivin et al., 1993; Marion et al., 1991, 1993). For example, in a study of *ras* gene mutations by allele specific oligonucleotide hybridization in tumors of VC-exposed rats, 1 of 5 (20%) ASLs was found to contain a $G \rightarrow A$ transition at the second base of codon 13 of the N-*ras*-A gene (Boivin et al., 1993).

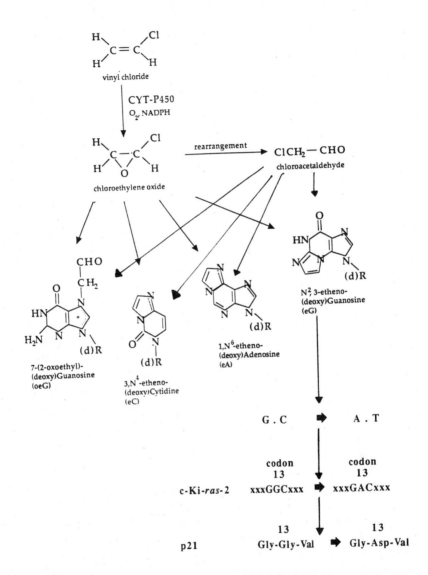

FIGURE 1. Proposed carcinogenic mechanism of vinyl chloride in the production of angiosarcomas of the liver.

In addition, in a similar study of *ras* gene mutations in tumors of VC-exposed workers, 15 of 18 (83%) ASLs examined to date have been found to contain a $G \rightarrow A$ transition at the second base of codon 13 of the c-Ki-*ras*-2 gene (Marion et al., 1991, 1993). These *ras* mutations at codon 13 ($GGC \rightarrow GAC$) are known to be capable of converting the gene into an actively transforming oncogene in model systems and, thus, may be causally related to the carcinogenic process (Bos et al., 1985; Barbacid, 1987). These *ras* mutations would result in the substitution of an aspartic acid for the normal glycine at amino acid residue 13 in the encoded p21 protein products (Barbacid, 1987). This amino acid substitution is believed to produce a conformational change in p21 which may be responsible for altering its intrinsic GTPase activity, thus affecting signal transduction within the cell (Pincus et al., 1985).

The altered Asp 13 p21 protein can be distinguished from normal p21 and other mutant p21s immunologically with mouse monoclonal antibodies that are specific for this mutant protein (La Vecchio et al., 1990). For example, Asp 13 p21 can be identified with these antibodies in cells by immunohistochemistry and in the cell lysate by immunoblotting of cells in culture known to contain the respective mutant *ras* gene (La Vecchio et al., 1990; DeVivo et al., 1994). Asp 13 p21 can also be identified with these antibodies by immunoblotting in the extra-cellular supernatant of such cells, suggesting that in an analogous situation *in vivo*, this protein may be similarly detectable in extra-cellular fluids such as serum (La Vecchio et al., 1990; DeVivo et al., 1994).

In other studies, increased total p21 protein or the presence of other mutant p21 proteins have been identified in the serum of mice bearing tumors that over-express the ras gene or contain mutant *ras* genes, respectively (Kakkanas and Spandidos, 1990; Hamer et al., 1991), and increased total p21 protein has been detected in the serum of humans with cancer or who are at-risk for the development of cancer due to their carcinogenic exposures (Brandt-Rauf et al., 1992). Similarly, detection of the Asp 13 mutant p21 in serum may be possible following VC exposure, and, thus, the ability to detect this Asp 13 p21 in serum may be a potential molecular epidemiologic biomarker for the study of this chemical carcinogenic process *in vivo* in exposed humans (Brandt-Rauf et al., 1994). The goal of this research was to explore that possibility.

METHODS

Based on job categories and years worked, 60 heavily VC-exposed workers employed in VC polymerization operations in France since 1950 were selected for study. These workers included 5 cases of ASL, 1 case of hepatocellular carcinoma (HCC), 9 cases of liver angiomas, and 45 workers with no liver neoplasia. These individuals were all white males with an average age of 57 years and with an average of 19.5 years of exposure to VC, including an average of 12.2 years of very high estimated exposure to VC prior to 1974 when strict exposure controls were established. Fresh frozen tumor tissue samples were available for analysis from the 6 cancer cases, and stored frozen serum samples were available for analysis from all 60 exposed workers. For 8 of the exposed workers, two separate serum samples that had been collected 1-2 years apart were available; and for one of the ASL cases, 7 separate serum samples that had been collected over the course of the disease were available. Six normal liver samples for use as tissue controls were available from age-race matched unexposed autopsy patients with non-cancer diagnoses, and 28 normal blood samples for use as serum controls were available from age-sex-race matched unexposed hospital patients with non-cancer diagnoses.

PCR-amplified DNA from tissue samples was analyzed by allele specific oligonucleotide hybridization for point mutations at codons 12, 13, and 61 of the c-Ha-*ras*, c-Ki-*ras*, and N-*ras* genes, as described (Marion et al., 1991). Tissue immunohistochemistry for Asp 13 c-Ki-*ras* p21 was performed with a primary mouse monoclonal antibody raised against a synthetic peptide corresponding to amino acid residues 5-16 of the p21 protein with Asp at position 13 and with a primary mouse monoclonal antibody raised against a synthetic peptide corresponding to amino acid residues 157-180 of the c-Ki-*ras* p21 protein, as described (DeVivo et al., 1994). Serum immunoblotting for Asp 13 c-Ki-*ras* p21 was performed with the same two primary mouse monoclonal antibodies as used for the tissue immunohistochemistry, as described (DeVivo et al., 1994). Tissue DNA analysis, tissue immunohistochemistry, and serum immunoblotting results were recorded blinded to the results of the other assays and blinded to diagnosis.

RESULTS AND DISCUSSION

Four of the five (80%) cases of ASL were found to contain the c-Ki-*ras* codon 13 mutation and to express the corresponding Asp 13 c-Ki-*ras* mutant p21 protein in their tumor tissue and in their serum (Figure 2). One case of ASL and the case of HCC found not to contain the mutation failed to express the mutant protein in their tumor tissue or serum (Table 1). These results suggested that detection of serum mutant p21 protein can be a valid surrogate for mutant *ras* gene expression at the tissue level *in vivo*.

Table 1   Results of DNA and protein analyses for Asp 13 c-Ki-*ras* in vinyl chloride-exposed workers and controls

	Patient	DNA with GAC at Ki-*ras* codon 13	Tissue Asp 13 Ki-*ras* p21	Serum Asp 13 Ki-*ras* p21
ASL	48	+	+	+
	73	+	+	+
	155	+	+	+
	562	+	+	+
	324	-	-	-
	Total	80% +	80% +	80% +
HCC	502	-	-	-
	Total	0% +	0% +	0% +
Angioma (n=9)	Total			89% +
Exposed (n=45)	Total			49% +
Control Sera (n=28)	Total			0% +
Control Tissue (n=6)	Total		0% +	

The serum results also appeared to be reasonably consistent over time or with the course of disease. For example, for the 8 pairs of serum samples, 6 were positive for Asp 13 p21 in both samples, one was negative in both samples, and one was negative in the initial sample and positive in the later sample which may be attributable to the occurrence of the mutation in the intervening years. In the ASL case with 7 serum

samples, relative quantitation of the intensity of p21 bands on the immunoblots by laser densitometry suggested that the level of the mutant
protein in serum paralleled the clinical course of the disease (Figure 3).

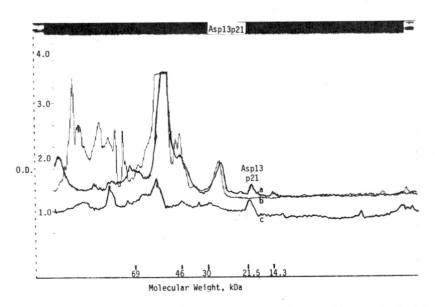

FIGURE 2. Densitometer tracing of representative immunoblots probed with
the primary mouse monoclonal antibody raised against a synthetic peptide corresponding to p21 residues 5-16 with Asp 13 of serum from vinyl chloride exposed worker with angiosarcoma of the liver known to contain the corresponding
mutant c-Ki-*ras* gene (a), serum from a matched normal healthy control (b),
and lysate from a positive control cell line known to express the Asp 13 c-Ki-*ras*
p21 protein (c). Note the presence of peaks at 21 kDa that are in (a) and (c)
but absent in (b).

In this case, after initial partial hepatectomy and chemotherapy as
attempted curative procedures, the serum mutant p21 was very low.
However, it began to increase and disease recurrence in the liver was
confirmed by computerized tomography. Radiotherapy was unsuccessful
and serum levels continued to increase. Successful chemoembolization
of the liver lesions plus interferon therapy caused a temporary drop
in serum levels, but these rebounded with disease progression as the
patient deteriorated clinically. Prior to death, the serum levels declined
again, which could have been due to a combination of calcification of

some of the liver segments with lesions and a pre-terminal decrease in overall protein synthesis by the liver.

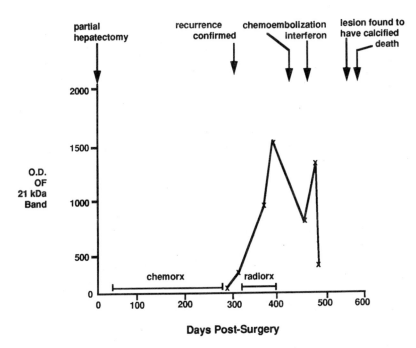

FIGURE 3. Time course of relative quantitation of serum Asp 13 p21 by volumetric integration of optical densities by laser densitometry of 21 kDa bands on immunoblots probed with the primary mouse monoclonal antibody specific for the Asp 13 p21 protein of serum samples obtained at various points in time following surgery for the primary tumor of a vinyl chloride-exposed worker with an angiosarcoma of the liver known to contain the Asp 13 c-Ki-ras mutation. The level of mutant protein appeared to parallel the clinical course of the disease.

The other serum results were also striking (Table 1). None of the controls was found to have detectable levels of the mutant p21 in their serum. However, 8 of 9 (89%) of the VC-exposed workers with angiomatous lesions of the liver were found to have serum positive for the Asp 13 p21. Liver angiomas have been associated with VC exposure in animal models, and hyperplasia and proliferation of sinusoidal lining cells with sinusoidal dilation have been associated with VC exposure in humans (Maltoni et al., 1984; ATSDR, 1993). Therefore, it is possible that the angiomatous lesions in these workers represent a potentially pre-malignant condition. To date, at least one non-dysplastic pre-angiosarcomatous liver lesion in a VC-exposed worker has been found to contain the characteristic codon 13 c-Ki-*ras* mutation (Marion et al., 1991). In addition, 22 of the 45 (49%) VC-exposed workers without neoplasia were found to contain the Asp 13 mutant p21 in their serum. These results suggest that the codon 13 mutational activation of c-Ki-*ras* may be an early event in VC carcinogenesis and that identification of the Asp 13 serum protein may be a potential biomarker of cancer risk in these individuals. Follow-up of the serum-positive and serum-negative workers in this cohort should provide confirmation of this.

To date, all serum-negative individuals remain healthy, but one of the serum-positive individuals has developed a liver lesion suspicious for ASL. It is important to note that in an analogous study of $Gly \rightarrow Asp$ DNA mutations at codon 13 of N-*ras* in leukemias and pre-leukemic myelodysplastic syndromes, cases of myelodysplasia that advanced to leukemia had the Asp 13 *ras* mutation, whereas cases without the mutation remained in the pre-leukemic state (Liu et al., 1987).

The serum results for all 60 exposed workers and 28 controls were also stratified by degree of VC exposure in terms of total years exposed by decades, and, using the unexposed controls as the reference, odds ratios were calculated for each stratum (Table 2).

Table 2   Dose-response relationship between the biomarkers, Asp 13 p21 and years of VC exposure.

	Biomarker (ASP 13 p21)	Biomarker (ASP 13 p21)	
Years of Exposure	Yes	No	Odds Ratio
0	0	28	1
<10	4	6	37
10-19	11	11	56
20-29	13	7	104
≥30	6	2	168

Chi Square = 24.986
$p < 10^{-5}$

The χ^2 value for the linear trend of serum mutant p21 positivity with increasing exposure was 25 with p< 0.00001. Since increasing VC exposure is thought to be associated with increasing cancer risk (Heldaas et al., 1984), this suggests that the presence of serum Asp 13 p21 may indeed be consistent with potentially elevated cancer risk in these individuals.

An important goal of molecular epidemiologic studies such as this one is disease prevention by using biological markers of presumed critical events in the disease pathway in order to better identify individuals at risk before clinical onset and thus to allow more effective intervention. Therefore, the potential identification of elevated cancer risk without appropriate therapy is of limited utility. However, we have found that the anti-tumor compound L-β-(5-hydroxy-2-pyridyl)-alanine (azatyrosine), which has been shown to be effective against other mutant *ras* transformed cells (Shindo-Okada et al., 1989; Izawa et al., 1992), causes permanent reversion of the Asp 13 c-Ki-*ras* containing HCT 116 human cancer cell line to a normal phenotype that has lost the ability to grow in soft agar or to produce tumors in nude mice (Brandt-Rauf et al., 1994). Thus, azatyrosine may serve as a prototype for chemoprophylaxis of mutant *ras*-associated tumors, and, through a combination of early identification of cancer risk via serum oncoprotein biomarkers and specific chemoprophylaxis, it may be feasible to prevent a significant proportion of VC-associated ASLs and other mutant *ras*-related cancers.

ACKNOWLEDGEMENTS

We thank Drs. C. Trepo, Y. Zhang, S. Kahn, Z. Yamaizumi, and S. Nishimura for their assistance with this project. This work was supported in part by the National Institute for Occupational Safety and Health (0H00076), the U.S. Environmental Protection Agency (R818 624), the National Institutes of Health (ES05948), the Lucille P. Markey Charitable Trust, the Association pour la Recherche sur le Cancer and le Groupment des Enterprises Françaises pour la Lutte contre le Cancer.

REFERENCES

ATSDR-Agency for Toxic Substances and Disease Registry. 1993. Toxicological profile for vinyl chloride. TP-92/20. U.S. Department of Health and Human Services, Public Health Service, ATSDR, Atlanta.

Barbacid, M. 1987. Annual Reviews of Biochemistry 56:779-827.

Barbin, A., and H. Bartsch. 1986. The role of cyclic nucleic acid adducts in carcinogenesis and mutagenesis. B. Singer, and H. Bartsch, eds. IARC: Lyon. Pp. 345-358.

Barbin, A., R. J. Laib, and H. Bartsch. 1985. Cancer Research 45:2440-2444.

Boivin, S., M.-J. Marion, O. Froment, J.-C. Contassot, and C. Trepo. 1993. Proceedings of the International Congress on Occupational Health 24:258.

Bos, J. L., D. Toksoz, C. J. Marshall, M. Verlaan de Vries, G. H. Veeneman, A. J. van der Eb, J. W. G. Janssen, and A. C. M. Steenvoorden. 1985. Nature 315:726-730.

Brandt-Rauf, P. W., S. Smith, K. Hemminki, H. Koskinen, H. Vainio, H. Niman, and J. Ford. 1992. International Journal on Cancer 50:881-885.

Brandt-Rauf, P. W., I. DeVivo, M.-J. Marion, and K. Hemminki. 1994. Journal of Occupational Medicine (in press).

Cheng, K. C., B. D. Preston, D. S. Cahill, M.K. Dosanjh, B. Singer, and L. A. Loeb. 1991. Proceedings of the National Academy of Sciences (USA) 88:9974-9978.

DeVivo, I., M.-J. Marion, S. J. Smith, W. P. Carney, and P. W. Brandt-Rauf. 1994. Cancer Causes Control (in press).

Hamer, P. J., J. LaVecchio, S. Ng, R. De Lellis, H. Wolfe, and W. P. Carney. 1991. Oncogene 6:1609-1615.

Heldaas, S. S., S. L. Langard, and A. Anderson. 1984. British Journal of Industrial Medicine 41:25-30.

Huberman, E., H. Bartsch, and L. Sachs. 1975. International Journal on Cancer 16:644-693.

Izawa, M., S. Takayama, N. Shindo-Okada, S. Doi, M. Kimura, M. Katasuki and S. Nishimura. 1992. Cancer Research 52:1628-1630.

Kakkanas, A., and D. A. Spandidos. 1990. *In Vivo* 4:115-120.

La Vecchio, J. A., P. J. Hamer, S. C. Ng, K. L. Trimpe, and W. P. Carney. 1990. Oncogene 5:1173-1178.

Liu, E., B. Hjelle, R. Morgan, F. Hecht, and J. M. Bishop. 1987. Nature 330:186-188.

Maltoni, C., G. Lefemine, A. Ciliberti, G. Cotti, and D. Carreti. 1984. Experimental research on vinyl chloride carcinogenesis. Princeton:Princeton Scientific Publishers.

Marion, M.-J., O. Froment, J.-C. Contassot, and C. Trepo. 1993. Proceedings of the International Congress on Occupational Health 24:175.

Marion, M.-J., O. Froment, and C. Trepo. 1991. Molecular Carcinogenesis 4:450-454.

McCann, J., V. Simmon, D. Streitwiesser, and B. N. Ames. 1975. Proceedings of the National Academy of Sciences (USA) 72:3190-3193.

Pincus, M. R., P. W. Brandt-Rauf, R. P. Carty, J. Lubowsky, and M. Avitable. 1985. Journal of Protein Chemistry 4:345-352.

Shindo-Okada, N., O. Makabe, H. Nagahara, and S. Nishimura. 1989. Molecular Carcinogenesis 2: 159-167.

Singer, B., S. J. Spengler and J. T. Kusmierek. 1988. Chemical Carcinogens: Activation Mechanisms, Structural and Electronic Factors and Reactivity. P. A. Politzer and F. Martin, eds. Elsevier:Amsterdam. Pp. 188-207.

Swenberg, J. A., N. Fedtke, F. Ciroussel, A. Barbin, and H. Bartsch. 1992. Carcinogenesis 13:727-729.

Biomarkers and Occupational Health: Progress and Perspectives. 1995. Pp. 174-193

Validation Studies for Monitoring of Workers Using Molecular Cytogenetics

Tore Straume and Joe N. Lucas

Methods to assess past clastogenic exposures are needed particularly to monitor workers who may have received substantial exposures, but for whom real-time monitoring data are either unreliable or not available. For example, many Department of Energy (DOE) radiation workers were exposed to neutrons during the 1940s through the 1960s when neutron dosimetry was in its infancy. Also, many individuals have received substantial radiation exposures in connection with accidents, nuclear weapons testing, human experimentation, the atom bombs dropped on Hiroshima and Nagasaki, and various medical radiological procedures. Reliable methods are also needed to help assess exposures to hazardous chemicals and other agents.

A number of biological markers developed to evaluate persons with known or suspected exposures to hazardous agents are available and listed in Table 1. These include markers of cytogenetic effects (Lucas et al., 1992a; Littlefield et al., 1990, 1992; Krepinsky and Heddle, 1983), somatic mutations (Albertini et al., 1982; Langlois et al., 1987; Mendelsohn, 1990; Straume et al., 1991; Akiyama et al., 1992), and adducts (Rothman et al., 1990; Turteltaub et al., 1993). It is noted that useful attributes, such as the applicability of the marker to all individuals in a population, availability of *in vitro* and experimental animal model systems, variability between individuals, stability with time post-exposure, and availability of supporting data, vary considerably among the assays. For example, GPA and HLA are applicable to only 50% of the population, and several other assays have not been sufficiently characterized to provide a reasonable estimate of their inter-individual variability.

A particularly important parameter for monitoring of workers and for use in dose reconstruction is the stability of the assay with time post-exposure. Two of the biomarkers (translocations and GPA) appear to exhibit lifetime persistence and, thus, could potentially be used to help reconstruct exposures that may have occurred many years or even decades previously. However, of these two biomarkers, only for translocations are there *in vitro* and animal models for use in dose-response characterization.

TABLE 1. A COMPARISON OF USEFUL ATTRIBUTES OF
SELECTED BIOMARKERS

Biomarker	Human in vivo	Human in vitro	Animal model	Inter-person variation[a]	Applicable time post-exposure[a]
Translocations[b]	yes	yes	yes	low	0-lifetime
Dicentrics	yes	yes	yes	low	0-6 mo
Micronuclei	yes	yes	yes	high	0-6 mo
HPRT[c]	yes	yes	yes	medium	1 mo-1 yr
GPA[d]	50%	no	no	high	6 mo-lifetime?
TCR[e]	yes	no	no	high	1 mo-2 yr
HLA[f]	50%	yes	no	?	1 mo-1 yr
SCEs	yes	yes	yes	?	0-6 mo
DNA adducts	yes	yes	yes	?	0-6 mo
Protein adducts	yes	yes	yes	?	0-6 mo

[a] Estimates based on studies cited in the text and evaluation by the authors.
[b] Reciprocal chromosome translocations.
[c] Hypoxanthine phosphoribosyltransferase assay.
[d] Glycophorin-A somatic mutation assay.
[e] T-cell antigen receptor mutation assay.
[f] Human leukocyte antigen mutation assay.

As recently as three years ago, it was doubtful whether any biomarker would be useful in the determination of radiation doses for workers exposed a long time ago or for those who received protracted low-level exposures over many years. However, our results for reciprocal translocations during the past couple of years have substantially demonstrated that this biomarker may be stable enough to reconstruct radiation doses for *individual* workers including those exposed *within* standard dose limits. In addition, this method may also be used to reconstruct dose for individuals exposed many years or even decades previously.

To actually perform such dose reconstruction requires a great deal of quantitative information, such as an accurate measurement of translocation stability, relevant dose-response information, and an assessment of the related uncertainties. This is particularly true for workers because the exposure rate is generally low, variable, and protracted over many years.

Molecular Cytogenetics

The "chromosome painting" technology employs fluorescence *in situ* hybridization (FISH) to rapidly and accurately detect chromosome abnormalities such as stable reciprocal translocations in human cells (Pinkel et al., 1986; Lucas et al., 1992a; Straume et al., 1992). In this technique, a small blood sample is obtained, the T-lymphocytes are cultured, and metaphase spreads are prepared on glass slides using standard cytogenetic methods (Evans et al., 1979). Pan-centromeric probes are used in combination with the FISH method to discriminate between reciprocal translocations and dicentrics (Lucas et al., 1992a; Straume and Lucas, 1993). FISH is generally performed using a cocktail of composite DNA probes specific for the largest chromosomes, i.e., numbers 1, 2, and 4, which comprise 22% of the genome and allow the detection of 35% of all translocations. The data obtained from only a few chromosomes using FISH are then scaled-up to assess full genome translocation frequencies. It has indeed been demonstrated that the frequency measured using FISH can be accurately scaled-up to full genomic frequency by assuming a random distribution of break points (Lucas et al., 1989a, 1989b, 1992a).

To visualize interchromosomal exchanges, metaphase spreads from lymphocytes are stained yellow with probes for selected chromosomes and blue with a pan-centromere probe. Non-target chromosomes are counterstained red with propidium iodide. Exchange aberrations are recognized as bi-color (part red and part yellow) chromosomes. The aberrations are thus scored as reciprocal translocations if the two derivative chromosomes each have one blue-stained centromere (Figure 1).

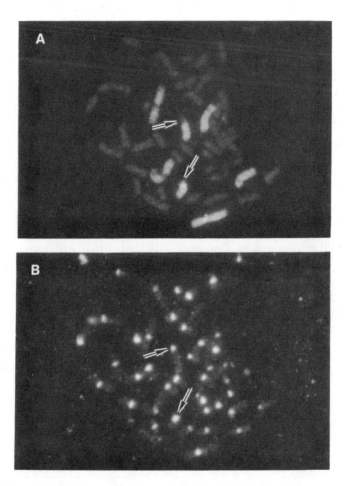

FIGURE 1. Photomicrographs showing FISH with whole-chromosome probes for chromosomes 1, 2, and 4 and a pan-centromere probe to human metaphase spreads. Whole chromosome-probe binding sites appear bright, unstained DNA appears darker, and the binding sites of the pan-centromere probes also appear bright. (a) Bright (FITC) and darker (PI) fluorescence from a metaphase spread carrying a translocation between chromosome 2 and a non-target chromosome (arrows). (b) Pan-centromeric probes showing bright (AMCA) fluorescence from the same metaphase spread. The two derivative chromosomes show only one centromere each (Color photographs are seen in Lucas et al., 1992a).

They are scored as dicentrics if the derivative chromosomes have two centromeres, and are scored as fragments if the derivative chromosomes have no centromeres. All aberrations seen using FISH are recorded on film for subsequent evaluation.

Validation of FISH for Cytogenetics

As indicated above, full genomic translocation frequencies are accurately obtained after painting only a small fraction of the genome. This finding was of critical importance because it permitted scaling to full genome equivalents from only a few painted chromosomes. Reciprocal translocation frequencies measured using FISH were compared with those measured using G-banding in blood lymphocytes from 20 Hiroshima atom bomb survivors (Lucas et al., 1992a). The frequencies measured using FISH for chromosomes 1, 2 and 4 were converted to full genome equivalents and then plotted against translocation frequencies measured by G-banding for all chromosomes for the same individuals (Figure 2). The results demonstrated that FISH provided reciprocal translocation frequencies that compared well with those measured by the standard G-banding method.

Although G-banding is universally accepted as an accurate method to detect chromosome translocations, it is far too labor intensive for biodosimetry applications. Thus, the demonstration that the much faster FISH method provided accurate results when scaled to the full genome, made it practical for reciprocal translocations to be considered as possible biomarkers of exposure and individual dose.

Persistence of Translocations

Historically, the use of cytogenetics in biological dosimetry has focused on dicentric chromosomes which are easy to observe using conventional methods (Bender et al., 1988). However, dicentric aberrations are unstable with time after exposure (Buckton et al., 1978; Littlefield et al., 1990) and thus are not useful for biological dosimetry more than a few months after exposure or for assessing chronic or multiple exposures.

Although translocations persist for many years after exposure, measurements that accurately quantify the stability of this biomarker have been made only recently. This is because the development of molecular cytogenetics made it feasible to score a large number of cells for reciprocal translocations so that good statistics can be obtained (Pinkel et al., 1986; Lucas et al., 1989a, 1992a, 1992b; Straume et al., 1992).

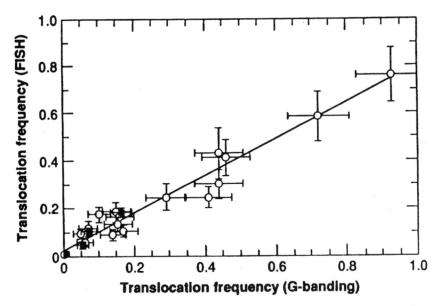

FIGURE 2. Comparison of reciprocal translocation frequencies measured using FISH and conventional G-banding techniques. Open circles: Translocation frequencies measured for Hiroshima A-bomb survivors using FISH for chromosomes 1, 2 and 4 were converted to genome equivalents and plotted against translocation frequencies measured by G-banding. Solid squares: Translocation frequencies measured for accidentally-exposed DOE workers using FISH for chromosomes 1, 3, and 4 were converted to genome equivalents and plotted against translocation frequencies measured for all chromosomes for the same workers by analysis of conventionally stained metaphase spreads. All error bars are 1 SD (Data from Lucas et al., 1992a).

The following is a summary of our results using molecular cytogenetics to quantify the persistence of reciprocal translocations in previously exposed individuals:

- We evaluated a worker in Switzerland who accidentally inhaled tritium oxide in 1985. The accident victim was a healthy 33-year-old woman who incorporated about 35 GBq of tritiated water. Dose was determined within a few weeks after the accident using both urinalysis and biodosimetry (Lloyd et al., 1986a). Biodosimetry involved measurement of the dicentric frequency in blood lymphocytes. Dosimetry results indicated that the woman had received 0.44 Gy of absorbed radiation dose to the soft tissues of the body (Lloyd et al., 1986a). Biodosimetry results obtained for this individual in 1991 (six years after exposure) using chromosome painting to measure reciprocal translocations (Lucas et al., 1992b) were identical to the dosimetry results obtained immediately after the accident from urinalysis and dicentrics. As shown in Table 2, the net frequency of reciprocal translocations measured in 1991 was the same as the frequency of dicentrics measured immediately after the accident. Because dicentrics and translocations are induced with the same frequency (Straume and Lucas, 1993), these results indicate that the translocation frequency for this individual has remained unchanged since exposure six years previously. The resultant dose estimates based on biodosimetry and urinalysis (also listed in Table 2) are in good agreement. This result was the first indication that the translocation frequency measured in human blood T-lymphocytes remains identical to its induction frequency many years after exposure, even though about 70% of the blood T-lymphocytes had been repopulated from stem cells during this study period.

TABLE 2. BIODOSIMETRY RESULTS FOR AN ACCIDENT
VICTIM EXPOSED TO TRITIATED WATER IN 1985

Year	Cells	Net exchanges per cell	Dose estimates (Gy)	
			Biodosimetric	Urinalysis
1985	1000	0.033±0.006	0.42[a]	0.44[b]
1991	1000	0.036±0.007	0.44[c]	0.47[d]

[a] Based on dicentrics measured 37 days post-exposure (Lloyd et al., 1986a).
[b] Based on urinalysis results obtained 37 days post-exposure (Lloyd et al., 1986a).
[c] Based on reciprocal translocations measured in 1991 using FISH.
[d] Dose as of 1991, based on urinalysis.

- A DOE worker exposed occupationally to radiation during the 1950s through 1970s was evaluated. His whole-body penetrating exposure was always within the DOE dose limits of 0.05 Sv per year. In 1989, the best-estimate dose obtained from the measured translocation frequency was 0.5±0.2 Sv, in good agreement with the total integrated dose recorded in his official dosimetry record from badge readings of 0.56±0.20 Sv. In addition to translocations, we used three other biomarkers to evaluate this individual: GPA, dicentrics, and micronuclei (Straume et al., 1992). GPA mutation frequencies, which have been shown to be persistent (Langlois et al., 1987), were also significantly elevated. However, dicentrics and micronuclei were not elevated above background levels as expected from their instability and the exposure pattern of this worker (Straume et al., 1992). Although this is only one individual, these results suggest that stable biomarkers can be detected in workers exposed within government established dose limits.

- In 1993 we evaluated five rhesus monkeys exposed to 2.3 GeV protons by the National Aeronautics and Space Administration (NASA) in the early 1960s. The biodosimetric doses were in good agreement with the doses actually delivered in 1963. Because 30 years in the lifespan of these primates is equivalent to about 90 years in the lifespan of humans, these measurements suggest a lifetime persistence of the reciprocal translocation frequency in the stem cells of blood lymphocytes. Also, these results showed very little inter-individual variation in either response or stability.

- Dose was also reconstructed for a Ukrainian scientist exposed in connection with the Chernobyl accident in 1986. Biodosimetry in 1993 indicated a dose of 0.33±0.12 Sv, which is essentially identical to the dose estimate of 0.3 Sv from independent physical measurements.

Table 3 summarizes studies that provide valuable information on the persistence of translocations in individuals. These studies were selected because measurements of translocation frequencies were made two or more times on the same individuals over several years permitting temporal comparisons to be made. Results indicate that there is no significant change in the translocation frequency during the time intervals evaluated. Note that prior to our recent results, there was a 17-year gap in the data between time of exposure and the first measurements.

TABLE 3. STUDIES THAT HAVE MEASURED TRANSLOCATION FREQUENCIES IN
THE SAME INDIVIDUAL(S) OVER A LONG PERIOD OF TIME POST-EXPOSURE

Study population	Individuals in population	Duration of follow-up (in years)	Source of data
Hiroshima/Nagasaki victims	> 100	23-40	Stram et al., 1993
Swiss tritium worker	1	0-6	Lucas et al., 1992b Lloyd et al., 1986a
Y-12 Accident victims	6	17-31	Littlefield et al., 1984 Lucas et al., 1992a
Chernobyl victim	1	0-7	Lucas & Straume et al. (unpublished data)
Rhesus monkeys	5	0-30	Lucas & Straume et al. (unpublished data)

In Vitro Calibration Curves

Calibration curves are required if a biomarker is to be used for
quantitative exposure and dose assessment. Such curves define the relationship between biomarker frequency and dose. For radiation-induced
chromosome aberrations, calibration curves are generally obtained using human lymphocytes exposed *in vitro*, which have been shown to be
identical to those exposed *in vivo* (Brewen and Gengozian, 1971). The
dose-response curve for acute exposure is linear-quadratic and the curve
for chronic exposure is linear. For a given radiation dose, both appear
to have identical linear slopes at low doses, i.e., the alpha coefficient.
This relationship is important because once the alpha coefficient has
been measured for a particular radiation type, it could then in principle
be used as a calibration curve for all low-level exposures involving that
radiation, without concern as to the actual dose rates and temporal
patterns of exposure.

The need for improved quantification of the alpha coefficient for low-LET radiations is illustrated by data in Figure 3. These data, which are for dicentric aberrations in human lymphocytes induced by radiation *in vitro*, illustrate two important points: (1) the alpha coefficient decreases substantially with increasing energy of common low-LET radiations and (2) large differences exist between alpha coefficients obtained by various laboratories, even though they were obtained using the same kinds of radiations.

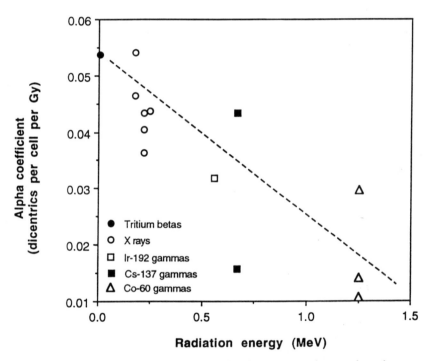

FIGURE 3. Alpha coefficients measured for dicentrics in human lymphocytes as a function of photon energy (from data in Todorov, 1975; Dufrain et al., 1980; Ziemba-Zoktowska et al., 1980; Takahashi et al., 1982; Prosser et al., 1983; Bauchinger et al., 1984; Fabry et al., 1985; Lloyd et al., 1986b; Littlefield et al., 1990.)

The dependence of alpha on photon energy (factors of 2 to 3 decrease between 200 kVp X-rays and 1.25 MeV ^{60}Co gamma rays) is too great to be ignored in biological dosimetry and must, therefore, be accurately measured *in vitro* for the radiations of concern. Also, the large uncertainties in the alpha coefficients themselves must be reduced so that an accurate relationship between alpha and radiation energy can be established.

For acute exposures below ~0.3 Gy and for all exposures at low to moderate dose rates, the alpha coefficient dominates the radiation-induced translocation frequency. Thus, for the low to moderate exposures of most practical concern in risk assessment, biodosimetry would not be possible without a well defined alpha coefficient.

The alpha coefficients now being measured in our program will provide many of the calibration curves necessary for radiation biological dosimetry using the molecular cytogenetics technology. Such data are particularly timely as molecular cytogenetics is becoming more generally employed in dose reconstruction.

Background Frequencies

The background frequency must be predictable within known uncertainty limits. To date, we have measured almost two dozen unexposed controls using FISH. These control subjects range from about 30 to 70 years of age (Table 4). The reciprocal translocation frequencies measured using FISH range from about 2 to 10 per 1,000 cells, which is in good agreement with frequencies reported in other studies for "unexposed" individuals (also listed in Table 4). Our highest frequencies were seen in older Hiroshima controls who were about 60 to 70 years of age. At the present time, our best-estimate background frequency for reciprocal translocations in adults obtained from the data in Table 4 is ~5 to 6±2 reciprocal translocations per 1,000 cells.

Although the background results for reciprocal translocations available from different study populations are in general agreement, further work is needed to quantify the effect of potentially important factors such as age and lifestyle.

Biodosimetry for Leukemia Patients

The inability to accurately measure pre-cancer induced genetic damage in the blood cells of patients with leukemia or lymphoma has prevented the use of available biodosimetric methods to determine prior exposure to clastogenic agents in these patients. This is because a substantial amount of genetic damage that is related to the disease itself appears in the blood cells of these patients and thus masks genetic damage that may have been present prior to the disease. We have developed a new approach that may be used to measure pre-cancer induced chromosomal aberrations in patients with B-cell leukemia by totally separating the T lymphocytes from the malignant B lymphocytes. The approach employs stable chromosome translocations and thus will detect prior exposures above the detection limit of ~0.05 to 0.1 Gy. The utility of this approach was illustrated using blood lymphocytes from a nuclear dockyard worker who claimed that his B-cell leukemia was induced by work-related radiation exposures (Lucas et al., 1994). Blood lymphocytes were obtained after diagnosis of the disease, but prior to therapy; measurements were made of the frequencies of chromosomal abnormalities in PHA-stimulated lymphocytes without prior separation of T and B cells and in T lymphocytes after complete separation from B-cells using a rosetting technique. Results indicated that the separation of T cells prior to PHA stimulation eliminates the cancer-related chromosomal damage and thus appears to facilitate biodosimetry of pre-cancer exposures in such patients.

Partial Body Exposures

For exposures other than whole body, the distribution of the dose within the body becomes important for the interpretation of the biomarker frequency. For example, inhalation or ingestion of radioiodine results in a dose primarily to the thyroid. Biodosimetry using blood lymphocytes would therefore not be appropriate in such cases. Another example of considerable practical importance is the radiation exposures to body extremities such as fingers or hands. Again, blood lymphocytes would not be useful because of the very small body volumes exposed. Development of biodosimetry assays for specific organs, such as thyroid and skin, would be particularly useful in assessing partial body exposures.

Chemical Workers

A study is under way to evaluate benzene workers through the use of chromosome painting. This work is in support of an epidemiology study conducted by the National Cancer Institute (NCI). Although only preliminary results are currently available, there seems to be extensive damage involving certain chromosomes and this observation will be further explored as a possible marker of exposure to benzene. Also, dose-response relationships for selected chemicals are being evaluated in our laboratory using chromosome painting. These dose-response studies are done using human lymphocytes exposed *in vitro* and include both known clastogens as well as certain food mutagens. Joint studies with NCI are also under way to evaluate primates exposed to food mutagens.

TABLE 4. SUMMARY OF RECIPROCAL TRANSLOCATION
FREQUENCIES IN CONTROL SUBJECTS

Subjects evaluated	Method	Number of subjects	Number of cells	Number of reciprocal translocations	Reciprocal translocations per 1000 cells
Hiroshima[a]	Size grouping	263	24,414	137	5.6
Hiroshima[b]	FISH	7	4,831	50	10.3
Oak Ridge[c]	Size grouping	81	16,215	63	3.9
China[d]	FISH	5	3,516	8	2.2
Croatia[d]	FISH	4	1,801	6	3.3
Oak Ridge[d]	FISH	1	1,502	5	3.3
Central Valley,[d] California	FISH	1	480	1	2.1
Livermore[d]	FISH	5	10,589	55	5.2
Total		367	63,348	325	

[a] Awa et al., 1978. Atom bomb survivors of mixed ages who received less than 0.01 Gy.
[b] Lucas et al., 1992a. Older atom bomb survivors (60 to 70 years of age) who received less than 0.01 Gy.
[c] Bender et al., 1988.
[d] Laboratory workers and scientists who have not received significant exposure to clastogenic agents.

Reducing Uncertainties

An essential requirement for the use of biomarkers in exposure assessment and dosimetry is quantification of the parameters that drive the overall uncertainty. In the simplest case (i.e., under conditions where αD defines the dose-response relationship), only three parameters must be known for biodosimetric reconstruction of radiation dose to an individual. These are the measured frequency in the individual (Yi), the background frequency (Yb), and the slope (α) of the calibration curve (Equation 1). Under such conditions, uncertainties in these three parameters will determine the overall uncertainty in the dose estimate.

$$D = \frac{Yi - Yb}{\alpha} \qquad (Equation\ 1)$$

Using actual values measured for a DOE radiation worker, the impact of reducing uncertainties in the α coefficient is illustrated. During the past couple of years, we have substantially reduced the uncertainty in the α coefficient for ^{60}Co gamma rays (which are similar in mean energy to the radiation received by the DOE worker). Listed in Table 5 are means, standard deviations (SDs), and assumed uncertainty distributions for the key parameters in Equation 1. Note that the α coefficient measured in 1992 was 0.023±0.016 and with additional measurements became 0.023±0.003 in 1994. This large reduction in the SD was the result of scoring almost 23,000 human lymphocytes irradiated with low doses of ^{60}Co gamma rays *in vitro*.

TABLE 5. PARAMETER VALUES AND ASSUMED UNCERTAINTY
DISTRIBUTIONS FOR A DOE RADIATION WORKER[a]

Parameter	Mean	SD	Distribution
Yi	0.017	0.006	Poisson
Yb	0.006	0.002	Lognormal
α (1992 value)	0.023	0.016	Lognormal
α (1994 value)	0.023	0.003	Lognormal

[a] See Equation 1.

The parameters in Table 5 were used in Equation 1 to calculate a best-estimate dose and uncertainty distributions for the exposed worker. The uncertainty distributions were modeled using a Monte Carlo computer code developed for such probabilistic analyses. The Monte Carlo code stochastically propagates the uncertainties for each parameter in Equation 1 and provides an uncertainty distribution that is consistent with the input data. The distribution in Figure 4a was calculated using the α coefficient available in 1992. The distribution in Figure 4b was calculated using our most recent α coefficient. Clearly, increasing the precision of only one of the key coefficients in Equation 1 has substantially reduced the uncertainty in the overall dose estimate for the exposed worker. A further comparison is shown in Table 6.

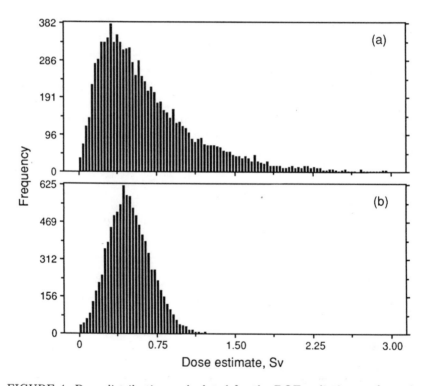

FIGURE 4. Dose distributions calculated for the DOE radiation worker using Monte Carlo sampling and the data in Table 5. (a) Distribution calculated using the alpha coefficient for reciprocal translocations available in 1992. (b) Distribution calculated using the improved alpha coefficient available in 1994.

TABLE 6. BIODOSIMETRY ESTIMATES AND UNCERTAINTIES
FOR A DOE RADIATION WORKER

Parameter	Using 1992 α value	Using 1994 α value	From personnel dosimeter
Mean (Yi)	0.69 Sv	0.49 Sv	0.56 Sv
SD	0.51 Sv	0.21 Sv	0.20 Sv
90% confidence interval	0.13-1.82	0.16-0.86	

These and related studies should provide the tools necessary to translate a measured translocation frequency into a meaningful estimate of exposure, dose, and risk. It is important to note that similar studies would be required before any biomarker can in fact be used reliably in quantitative exposure assessment.

ACKNOWLEDGEMENTS

This work was performed under the auspices of the U. S. Department of Energy by the Lawrence Livermore National Laboratory under contract number W-7405-Eng-48. We are especially grateful for the support of Dr. Harry Pettengill, DOE/EH.

REFERENCES

Akiyama, M., Y. Kusunoki, U. Shigeko, H. Yuko, N. Nakamura, and S. Kyoizumi. 1992. Evaluation of four somatic mutation assays as biological dosimeters in humans. Pp.177-182 in: Radiation Research: A Twentieth-Century Perspective. Academic Press, Inc., San Diego.

Albertini, R. J., K. S. Castle, and W. R. Borcherding. 1982. T-cell cloning to detect the mutant 6-thioguanine-resistant lymphocytes present in human peripheral blood. Proceedings of the National Academy of Sciences (USA) 79:6617-6621.

Awa, A. A., T. Sofuni, T. Honda, T. Itch, S. Neriishi, and K. Otake. 1978. Relationship between the radiation dose and chromosome aberrations in atomic bomb survivors of Hiroshima and Nagasaki. Journal of Radiation Research 19:126-140.

Bauchinger, M., L. Koester, E. Schmid, J. Dresp, and S. Streng. 1984. Chromosome aberrations in human lymphocytes induced by fission neutrons. International Journal of Radiation Biology 45:449-457.

Bender, M. A., A. A. Awa, A. L. Brooks, H. J. Evans, P. G. Groer, L. G. Littlefield, C. Pereira, R. J. Preston, and B. W. Wacholz. 1988. Current status of cytogenetic procedures to detect and quantify previous exposures to radiation. Mutation Research 196:103-159.

Brewen, J. G., and N. Gengozian. 1971. Radiation-induced human chromosome aberrations. II. Human *in vitro* irradiation compared to *in vitro* and *in vivo* irradiation of marmoset leukocytes. Mutation Research 13:383-391.

Buckton, K. E., G. Hamilton, L. Paton, and A. Langlands. 1978. Chromosome aberrations in irradiated ankylosing spondylitis. Pp. 142-150 in: Mutagen-induced chromosome damage in man. University Press, London.

Dufrain, R. J., L. G. Littlefield, E. E. Joiner, and E. L. Frome. 1980. *In vitro* human cytogenetic dose-response systems. P. 357 in: The Medical Basis for Radiation Accident Preparedness. K. F. Hubner and S. A. Fry, eds. Elsevier Science Publishing Company, New York.

Evans, H. C., K. E. Buckton, G. E. Hamilton, and A. Carothers. 1979. Radiation -induced chromosome aberrations in nuclear-dockyard workers. Nature 277:531-534.

Fabry, L., A. Leonard, and P. Wambersie. 1985. Induction of chromosome aberrations in G_o human lymphocytes by low doses of ionizing radiations of different quality. Radiation Research 103:122-134.

Krepinsky, A. B., and J. A. Heddle. 1983. Micronuclei as a rapid and inexpensive measure of radiation-induced chromosomal aberrations. Pp. 93-109 in: Radiation-induced chromosome damage in man. T. Ishihara and M. S. Sasaki, eds. Alan R. Liss, Inc., New York.

Langlois, R. G., W. L. Bigbee, S. Kyoizumi, N. Nakamura, M. A. Bean, M. Akiyama, and R.H. Jensen. 1987. Evidence for increased somatic cell mutations at the glycophorin A locus in atomic bomb survivors. Science 236:445-448.

Littlefield, L. G., A. M. Sayer, S. Colyer, E. E. Joiner, J. Outlaw, and L. Dry. 1984. Persistent radiation-induced chromosome lesions in lymphocytes of the Y-12 accident survivors: Evaluation 25 years post-exposure. Mammalian Chromosomes Newsletter 25:18.

Littlefield, L. G., E. E. Joiner, and K.F. Hubner. 1990. Cytogenetic techniques in biological dosimetry: Overview and example of dose estimation in 10 persons exposed to gamma radiation in the 1984 Mexican ^{60}Co accident. Pp. 109-126 in: Medical Management of Radiation Accidents, F.A. Mettler, Jr., C.A. Kelsey, and R.C. Ricks, eds. CRC Press, Inc., Boca Raton, FL.

Littlefield, L. G., A. M. Sayer, F. Sallam, E. L. Frome, and R.A. Kleinerman. 1992. Experience with the "cytochalasin-B method" for quantifying radiation-induced micronuclei in human lymphocytes. Pp. 183-188 in: Radiation research: A Twentieth-Century Perspective. Academic Press, Inc., San Diego.

Lloyd, D. C., A. A. Edwards, J. S. Prosser, A. Auf der Maur, A. Etzweiler, U. Weickhardt, U. Gossi, L. Geiger, U. Noelpp, and H. Rossler. 1986a. Accidental intake of tritiated water: A report of two cases. Radiation Protection Dosimetry 15:191-196.

Lloyd, D. C., A. A. Edwards, and J. S. Prosser. 1986b. Chromosome aberrations induced in human lymphocytes by *in vitro* acute X and gamma radiation. Radiation Protection Dosimetry 15:83-88.

Lucas, J., T. Tenjin, T. Straume, D. Pinkel, D. Moore, M. Litt, and J. Gray. 1989a. Rapid determination of human chromosome translocation frequency using a pair of chromosome-specific DNA probes. International Journal of Radiation Biology 56:35-44.

Lucas, J., T. Tenjin, T. Straume, D. Pinkel, D. Moore, M. Litt, and J. Gray. 1989b. Letter to the Editor. International Journal of Radiation Biology 56:201.

Lucas, J. N., A. Awa, T. Straume, M. Poggensee, Y. Kodama, M. Nakano, K. Ohtaki, U. Weier, D. Pinkel, J. Gray, and G. Littlefield. 1992a. Rapid translocation frequency analysis in humans decades after exposure to ionizing radiation. International Journal of Radiation Biology 62:53-63.

Lucas, J. N., M. Poggensee, and T. Straume. 1992b. The persistence of chromosome translocations in a radiation worker accidentally exposed to tritium. Cytogenetics Cell Genetics 60:255-256.

Lucas, J. N., G. Swansbury, R. Clutterbuck, F. Hill, C. Burk, and T. Straume. 1994. Discrimination between leukemia- and non-leukemia-induced chromosomal abnormalities in the patient's lymphocytes. International Journal of Radiation Biology (in press).

Mendelsohn, M. L. 1990. New approaches for biological monitoring of radiation workers. Health Physics 59:23-28.

Pinkel, D., T. Straume, and J.W. Gray. 1986. Cytogenetic analysis using quantitative, high sensitivity, fluorescence hybridization. Proceedings of the National Academy of Sciences (USA) 83:2934-2938.

Prosser, J. S., D. C. Lloyd, and A. A. Edwards. 1983. The induction of chromosome aberrations in human lymphocytes by exposure to tritiated water *in vitro*. Radiation Protection Dosimetry 4:21-26.

Rothman N., M. C. Poirier, M. E. Baser, J. A. Hansen, C. Gentile, E. D. Bowman, and P. T. Strickland. 1990. Formation of polycyclic aromatic hydrocarbon -DNA adducts in peripheral white blood cells during consumption of charcoal-broiled beef. Carcinogenesis 11:1241-1243.

Stram, D. O., R. Sposto, D. Preston, S. Abrahamson, T. Honda, and A. A. Awa. 1993. Stable chromosome aberrations among A-bomb survivors: An update. Radiation Research 136:29-36.

Straume, T., R. G. Langlois, J. Lucas, R. H. Jensen, W. L. Bigbee, A. T. Ramalho, and C. E. Brandao-Mello. 1991. Novel biodosimetry methods applied to the victims of the Goiania accident. Health Physics 60:71-76.

Straume, T., J. N. Lucas, J. D. Tucker, W. L. Bigbee, and R. G. Langlois. 1992. Biodosimetry for a radiation worker using multiple assays. Health Physics 62:122-130.

Straume, T., and J. N. Lucas. 1993. A comparison of the yields of translocations and dicentrics measured using fluorescence *in situ* hybridization. International Journal of Radiation Biology 64:185-187.

Takahashi, E., M. Hirai, I. Tobari, T. Utsugi, and S. Nakai. 1982. Radiation-induced chromosome aberrations in lymphocytes from man and crab-eating monkey. The dose-response relationships at low doses. Mutation Research 94:115-123.

Todorov, S. L. 1975. Radiation-induced chromosome aberrations in human peripheral lymphocytes. Exposure to X-rays or protons. Strahlentherapie 149:197-204.

Turteltaub, K.W. 1993. Studies on DNA adduction with heterocyclic amines by accelerator mass spectrometry: A new technique for tracing isotope-labelled DNA adduction. Pp. 293-301 in: Postlabelling Methods for Detection of DNA Adducts: D.H. Phillips, M. Castegnaro and H. Bartsch, eds. International Agency for Research on Cancer, Lyon, France.

Ziemba-Zoktowaka, B., E. Bocian, O. Rosiek, and J. Sablinski. 1980. Chromosome aberrations induced by low doses of X-rays in human lymphocytes *in vitro*. International Journal of Radiation Biology 37:231-236.

Biomarkers and Occupational Health: Progress and Perspectives. 1995. Pp. 194-214

Molecular Cytogenetic Approaches to the Development of Biomarkers

Joe W. Gray, Dan Moore, James Piper, and Ronald Jensen

The United States and the world are paying increasing attention to assessment of the deleterious effects of radiation and other environmental contaminants. These efforts are being driven by needs on several fronts. One important force is the congressional mandate to assess exposure in current and former employees of the United States Department of Energy (DOE). This effort will be increasingly important as site remediation activities sponsored by DOE accelerate and bring increasing numbers of workers into potentially toxic environments. Another important force is the need to learn as much as possible about the health effects of environmental contaminants through the study of exposed populations such as Chernobyl accident victims, the atomic bomb survivors in Japan, workers exposed in foundries (Perera et al., 1993), coke ovens (Motykiewicz et al., 1992), styrene (Wogan, 1992) and butadiene (Cowles et al., 1994) manufacturing. Efforts to accomplish this through physical monitoring are often unsuccessful since the agents may be unknown, and if known, may not be detectable using existing technical procedures. In addition, physical monitoring is difficult or impossible after exposure when monitoring equipment was not in place. An alternative is to measure the level of exposure by measuring the biological changes induced by the agents rather than the agents themselves. Such biomonitoring has been applied effectively for assessing the effects of radiation exposure (Bender et al., 1988). It is clear that techniques are needed to assess the exposure and/or the risk of genetic disease associated with a broad range of contaminants. However, it seems unrealistic to believe that a single assay will be able to accomplish this. Thus, in this paper we will concentrate on the assessment of radiation-induced genetic damage.

Work conducted over more than three decades has clearly shown that the frequency of chromosome aberrations increases with radiation dose, that the risk of cancer increases with increasing aberration frequency, and that the frequency of stable chromosome aberrations, on average, remains constant with time after exposure. Thus, the frequency of stable chromosome aberrations performs well as a biomarker of acute exposure to whole body radiation for a long time after exposure (Littlefield et al., 1991). The assay, however, has several important limitations (Littlefield et al., 1990): (a) The frequency of aberrations produced by occupational levels of radiation is low (typically $< 10^{-2}$ aberrations/cell) so that measurement of the frequency of aberrations is time-consuming and expensive (Bender et al., 1988). (b) The frequency of cells carrying translocations may fluctuate as a result of antigenic stimulation of lymphocyte subpopulations so that dose estimates based on cytogenetic aberrations in peripheral blood lymphocytes may fluctuate as well. (c) The baseline translocation frequency may vary among individuals, presumably as a result of variable exposure to clastogenic agents. (d) Radiosensitivity may vary among individuals. (e) The relationship between translocation frequency and level of exposure in humans is not well known for all radiation exposure conditions. This paper examines the extent to which these limitations can be eased or at least studied effectively through application of fluorescence *in situ* hybridization (FISH) (Pinkel et al., 1988) with whole chromosome probes (WCPs) (Collins et al., 1991) and via automated digital imaging microscopy (Piper and Granum, 1989). In addition, we assess the eventual utility of chromosome-based radiation dosimetry analysis assuming that these techniques have been fully implemented.

Another approach to genetic damage analysis is to assess exposure by measuring the mutation frequency at specific loci. Substantial work has now been performed on analysis of mutations involving glycophorin A (GPA) (Jensen et al., 1991), hypoxanthine phosphoribosyltransferase (HPRT) (Albertini et al., 1993), human leukocyte antigens (HLA) (Grist et al., 1992) and the T cell receptor (TCR) (Kyoizumi et al., 1992). Frequencies of mutations involving HPRT, HLA and TCR are not significantly elevated in the Japanese A-bomb survivors decades after exposure (Akiyama et al., 1992) and thus do not seem to be good long term radiation dosimeters. This may be due to selection against cells carrying these mutations. However, the frequency of red blood cells carrying mutations in GPA does seem to reflect exposure at long times after exposure (Langlois et al., 1993). Interestingly, on an individual basis, the frequency of cells carrying GPA mutations is not closely correlated with the frequency of cells carrying stable chromosome translocations. This suggests that it may be necessary to assess damage at several loci in order to accurately assess the biological effect of DNA damaging agents. Ideally, the loci tested will be those that are closely associated with untoward biological events such as cancer induction. We discuss in this paper the possibility of using comparative genomic hybridization (CGH) (Kallioniemi et al., 1992) to facilitate identification of radiation-specific aberrations that lead to cancer with the idea that these can form the basis for new bioassays of exposure.

ABERRANT CHROMOSOME ANALYSIS

The introduction of FISH with whole chromosome probes (WCPs) has changed biological dosimetry based on analysis of structural chromosome aberrations by allowing rapid detection of structural aberrations between chromosomes. This is important since two major applications of chromosome-based radiation dosimetry are to assess average exposure in occupationally or accidentally exposed populations and to assess individual exposure. The dose to most individuals from either occupational or accidental exposure will be small (typically < 250 mSv). The translocation frequency produced by such small doses is correspondingly small ($< 2 \times 10^{-2}$ translocations per cell at 250 mSv), and the background translocation frequency may be relatively high (low at birth and increasing with age to $\sim 2 \times 10^{-2}$ by age 70 [Tucker et al., 1994]), and may vary between individuals. Thus, the challenge is to be able to score a sufficient number of cells to allow statistically significant assessment of dose above background and to have an accurate assessment of the background translocation frequency. In FISH with WCPs, metaphase spreads are prepared for analysis using standard cytogenetic techniques, are denatured thermally to expose single stranded DNA and are "stained" by hybridizing with fluorescently labeled nucleic acid probes that contain sequences distributed at numerous sites along one or more target chromosomes (Pinkel et al., 1988; Trask and Pinkel, 1990). The probes are labeled so that they fluoresce at one wavelength (e.g., green if the fluorochrome is fluorescein) and all chromosomes are stained so they fluoresce at a different wavelength (e.g., red if the fluorochrome is propidium iodide). Figure 1a demonstrates the staining patterns that would occur in a metaphase spread carrying translocations or dicentric chromosomes that formed as a result of an exchange between a target and non-target chromosome. Figure 1b presents a photomicrograph of a metaphase spread containing a translocation after FISH with a WCP to one of the involved chromosomes. These can be scored quickly by a skilled analyst.

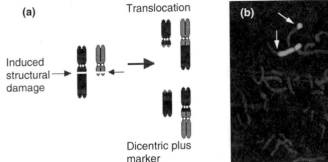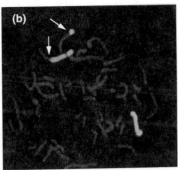

FIGURE 1. Panel a. Schematic representation of exchanges between a chromosome stained using FISH and a chromosome stained using a DNA counterstain that fluoresces at a different wavelength. The two breaks are resolved to form translocations or a dicentric plus acentric fragment. Panel b. Photomicrograph showing a translocation between two different chromosome types, one stained using FISH (light color) and one stained using a DNA specific counterstain (dark color). The translocation breakpoints are shown with arrows.

The choice of target chromosome type is open since WCPs are available for all human chromosomes (Collins, 1991). Some chromosome morphology is lost during FISH so that discrimination between chromosomes with 0, 1 or >1 centromeres (i.e., between stable and unstable aberrations) is difficult. This limitation can be overcome by staining the centromeres a third color (e.g., blue if the fluorochrome is amino methyl coumarin) using FISH with a probe that binds to all chromosome centromeres (Weier et al., 1991) so that unstable aberrations can be scored separately for dosimetry at short times after exposure or ignored if the unstable aberration frequency has decayed sufficiently to be an unreliable dosimeter. It is important to note that this assay detects only a fraction of the structural aberrations (exchanges between similarly stained chromosome types are invisible) and assumes that this is representative of the total. The fraction of total translocations, T_f, that can be detected is:

$$T_f = 2.05 f_p (1 - f_p) \qquad (Equation\ 1)$$

Where f_p is the fraction of the genome that is covered by the WCP
(i.e., that is painted [Lucas et al., 1992]); assuming that the translocations are randomly distributed throughout the genome. The accuracy of this important assumption has been questioned (Knehr et al.,
1994). However, the frequency of structural aberrations does increase
more-or-less linearly with chromosome size as expected for a random
distribution (Lucas et al., 1992). More importantly, as shown in Figure
2 the frequency of translocations estimated using FISH with WCPs and
converted to total genome translocation frequency using Equation 1 was
within 20% of that measured using G-banding in 25 Japanese A-bomb
survivors (Lucas et al., 1992).

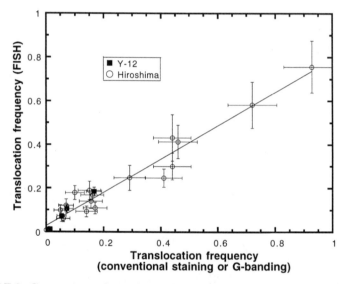

FIGURE 2. Comparison of translocations measured using G-banding or giemsa
staining with those measured after staining for chromosomes 1, 2, and 4 using
FISH (Lucas et al., 1992). Each point represents the translocation frequency
of a single individual. Open circles show frequencies for individuals exposed at
Hiroshima, and filled squares show frequencies for individuals exposed at the
Y-12 radiation accident. The translocation frequencies measured using FISH
were converted to whole genome equivalents using Equation 1.

A significant advantage of FISH with WCPs is that structural aberrations are sufficiently distinctive that they can be scored rapidly in poor

quality metaphase spreads and even in prematurely condensed chromo-somes (Evans et al., 1991).

FISH with WCPs has been applied in several studies assessing ra-diation exposure levels. Studied populations include the Japanese A-bomb survivors (Lucas et al., 1992; Nakano et al., 1993) and victims of the Goiania, Brazil ^{137}Cs accident (Ramalho et al., 1991). In addition, there are several unpublished studies involving Chernobyl accident vic-tims, British Nuclear Fuels Reprocessing Plant workers, and X-ray tech-nicians from U.S. hospitals (Jensen, personal communication). Several studies have been performed to describe the dose-response of chromo-some translocations measured by FISH with WCPs *in vitro* (Natarajan et al., 1992; Nakano et al., 1993; Schmid et al., 1992; Matsuoka et al., 1994). However, the technology for structural chromosome aberration-based dosimetry and the scientific basis for the assay require substantial development before widespread application is technically feasible or ap-propriate.

Dose Estimation

The utility of FISH with WCPs for biological dosimetry comes from the substantially increased speed with which metaphase spreads can be scored for structural aberrations. However, scoring sufficient metaphase spreads for accurate dosimetry is still a formidable task. We illustrate this by showing the relationship between the number of metaphase spreads scored and the magnitude of the 90% confidence in-terval (CI) for dose. To do this, we assume that: (a) the dose response is linear-quadratic with known coefficients and the response at zero dose is determined by the age and/or previous exposure history of the indi-vidual and (b) the responses are subject to Poisson variation. If y is the number of translocations detected by scoring n cells, assumptions a and b imply that y will have a Poisson distribution with expected value $n(\beta_0 + \beta_1 d + \beta_2 d^2)$, where d is the dose to be estimated. A point estimate for the dose is obtained by solving the quadratic equation for dose d, as shown in Equation 2:

$$\beta_2 d^2 + \beta_1 d + \beta_0 - \frac{y}{n} = 0 \qquad (Equation\ 2)$$

A 90% confidence interval for estimated dose can be obtained as illustrated in Figure 3.

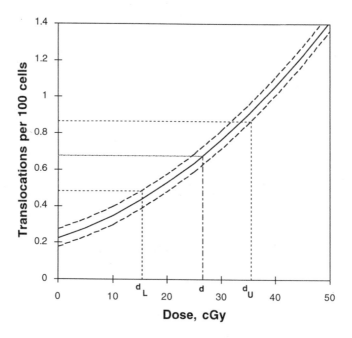

FIGURE 3. Method for determining 90% confidence interval (CI) for dose based on a background count of y_0=56 translocations based on scoring $n_0 = 25{,}000$ cells and a response of $y = 34$ translocations based on scoring 5,000 cells.

First, the upper and lower 90% confidence limits for the "true" translocation rate are determined. A simple approximation for these limits is $y \pm z_{0.95}\sqrt{y}$, where $z_{0.95} = 1.96$ is the standard normal deviate for the upper 5% of the standard normal distribution. When the linear (β_1) and quadratic (β_2) coefficients are known, the confidence bounds for the dose-response curve depend only on the 90% confidence limits for β_0. If β_0 is based on finding y_0 translocations when scoring n_0 cells, 90% confidence bounds for the dose-response curve are given by Equation 3:

$$\beta_2 d^2 + \beta_1 d + \frac{y_0}{n_0} \pm z_{0.95}\frac{\sqrt{y_0}}{n_0} \qquad (Equation\ 3)$$

The lower 90% confidence limit for the estimated dose is found by projecting the intersection of the lower limit for y and the upper limit for the dose-response curve onto the x-axis. Similarly, the upper 90% confidence limit for dose is found by projecting the intersection of the upper limit for y and the lower limit for the dose-response curve onto the x-axis.

The number of cells that must be scored to estimate dose to within a given confidence interval depends on the precision with which the dose response curve and the baseline translocation frequency are known. Table 1 shows the lower and upper 90% confidence limits for dose based on dose response curve coefficients described in the literature (Schmid et al., 1992) and a scoring strategy wherein the baseline (pre-exposure) translocation frequency is estimated with greater precision than the post-exposure frequency (i.e., more cells are scored to determine the baseline translocation frequency as might be the case where the baseline frequency is estimated from the combined results of analyzing many different individuals). These calculations assume a best case situation where the uncertainty in the dose response curve is determined only by the uncertainty with which the zero-dose translocation frequency is known. It is clear that tens of thousands of cells must be scored pre- and post-exposure in order to detect the effects of exposure to a doubling dose (0.16 Sv) or a tripling dose (0.26 Sv) with 90% confidence and that precise estimation of dose requires far more. For example, scoring all translocations in 25,000 metaphase spreads to estimate background and 5,000 metaphase spreads after exposure is barely sufficient to detect the effect of exposure to a doubling dose (0.16 Sv). Scoring these numbers of cells yields a very imprecise estimate of dose (the 90% confidence interval is expected to range from 0.02 to 0.26 Sv). These numbers assume that the pre- and post-exposure measurements were made on the same person and, therefore, do not take into account additional sources of variation such as interpersonal variation which has been estimated to have variance 2.25 as large as Poisson variation (Moore, unpublished). A variance multiple of 2.25 is the same as a 1.5 multiple of the standard deviation so that 50% more cells would have to be scored to compensate for inter-person variation. In addition, FISH with WCPs for chromosomes 1-4 allows detection of about 40% of all translocations so that 2.5 times more cells would have to be scored using FISH. In short, almost 100,000 cells (pre- and post-exposure) must be scored using FISH in order to be able to assess exposures in the range of a doubling dose.

TABLE 1. THE NUMBER OF CELLS THAT MUST BE SCORED TO
ESTIMATE A RADIATION DOSE

| No. of cells scored | | d=16 | | d=26.5 | |
Background	Exposed	Lower	Upper	Lower	Upper
10000	500	0	39	0	49
	1000	0	34	0	44
	2500	0	30	9	39
	5000	0	27	14	37
25000	500	0	38	0	48
	1000	0	33	0	43
	2500	0	28	11	38
	5000	2	26	15	35
50000	500	0	37	0	47
	1000	0	32	0	42
	2500	0	28	11	38
	5000	3	22	16	35

Lower and upper 90% confidence limits for dose based on scoring numbers of cells
indicated in the first two columns. The third and fourth columns are for exposure
to dose (d) of 16 cSv (a doubling dose); the fifth and sixth columns for exposure to
a dose of 26.5 cSv (a tripling dose).

Scoring rate

Scoring sufficient cells for exposure detection and dose estimation
in the < 0.3 Sv range is clearly impractical using banding analysis and
is difficult but possible using FISH. We estimate that a skilled analyst can find and score about 4 metaphases/min (i.e. approximately
1,400 metaphases per 6 hour day) after staining using FISH with WCPs
with most time devoted to metaphase finding. In our laboratory, automated metaphase finding has increased the scoring rate approximately
3-fold (about 4,000/analyst/6 hour day). In this approach, slides are
scanned automatically for metaphases and individual spreads are relo-
cated upon operator request for visual scoring. The rate limiting step in
this system is the time required for visual scoring since the system finds
metaphases at the rate of about 0.5/second (similar to that reported for
other metaphase finders). Another 5-fold increase in speed (20,000/an-

alyst/6 hour day) might be achieved by computationally eliminating clearly normal metaphase spreads prior to visual analysis (Piper, personal communication). Thus, straightforward improvements in technology will make analyses of low dose effects possible (e.g., 5 person-days per 10^5 cells with automation versus ~70 person-days per 10^5 cells with manual scoring). However, estimates of radiation dose from these data must be based on assumptions about background translocation frequencies, dose response relationships and hemopoietic stem cell biology that are not well understood at present.

Background Frequency

Detection of exposure above background and/or accurate estimation of dose at the low doses typically found in occupationally exposed workers or in individuals near accident sites requires that the background frequency be known accurately. Figure 4, from the recent work of Tucker et al. (1994) shows that this is not a simple matter, since the background frequency increases significantly with age and may vary among individuals of the same age; especially if they have different lifetime exposures to agents that induce translocations.

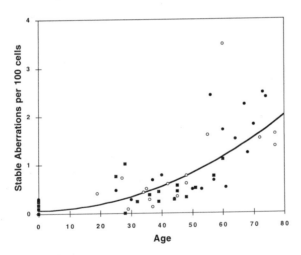

FIGURE 4. Increase in the background frequency of translocations as a function of age for 53 healthy individuals (Tucker et al., 1994). Open circles are for nonsmoking women, open squares for nonsmoking men; filled circles are for smoking women and filled squares for smoking men. The solid line is the least-squares regression fit to the data (y = 0.0615 + 0.000304 age^2).

Thus, assessment of exposure in large populations will require correction for age related increases in background frequency and may require assessment of baseline translocation frequencies in populations that live in similar environments. Assessment of individual exposure at very low doses may require assessment of pre-exposure translocation frequency in that individual (Tucker et al., 1994). In short, high precision analysis of background translocation frequencies in unexposed individuals and/or populations is essential before attempting routine assessment of exposure using chromosome translocation analysis. The magnitude of inter-individual variation is still not well known and variation of baseline translocation frequencies among different ethnic groups and between groups living in different environments has not been explored at all. Thus, additional studies of baseline translocation frequencies are critically needed to support assessment of low, occupationally relevant exposures to ionizing radiation.

Hemopoietic Effects

Estimation of dose from chromosome translocation frequencies measured in peripheral blood lymphocytes presumes a knowledge of stem cell biology that does not exist at this time. Specifically, it assumes that the translocations in the lymphocyte population sampled at the time of analysis are a democratic representation of the translocations induced in a precursor population and that the anatomic location of the precursor cells at the time of exposure is known. In fact, it is still not clear whether peripheral lymphocytes in adults derive from hemopoietic stem cells or from memory T-cells. Dose estimates may be in error if corrections for anatomic shielding are improper. Dose estimation also may be complicated by the fact that subpopulations of lymphocytes may proliferate preferentially in response to antigenic challenge. The extent to which this occurs is presently unknown. The complications arising from use of lymphocytes as the indicator population may be overcome in part by assessing translocation frequencies in other accessible cell populations such as peripheral blood granulocytes or skin epithelial cells that do not respond to antigenic challenge and that have better known and shorter lifetimes. This appears possible by measuring translocation frequencies in prematurely condensed chromosomes prepared from these cells (Evans et al., 1991).

Dose Rate Effects

Accurate dosimetry requires precise knowledge of dose response relationship. This is moderately well known for lymphocytes irradiated *in vitro* with acute doses of low and high LET radiation (Bender et al., 1988). However, there is evidence that the curves for acute exposure may not hold for long term environmental exposure (e.g., for nuclear plant workers or in the population surrounding Chernobyl). Specifically, Wolff and colleagues have demonstrated the existence of repair systems in human lymphocytes that can be induced by exposure to low doses of ionizing radiation (Wolff et al., 1990; Wolff et al., 1991; Wolff et al., 1993). Induction of these repair systems as a result of chronic exposure may substantially alter the shape of the dose-response curves. The impact of inducible repair in chronically exposed individuals on the shape of the response curve should be investigated carefully before dose estimation based on analysis of translocation frequencies in lymphocytes is attempted. Radiation dosimetry based on analysis of translocation frequencies also presumes no variation in radiosensitivity among individuals. However, studies of specific locus mutation frequencies in exposed populations have revealed some apparently normal individuals with extremely high mutation frequencies and who thus may be more radiosensitive than the average. Studies of translocation frequencies in the Japanese A-bomb survivors (Lucas et al., 1992) also deviate substantially from the frequencies expected based on estimates of physical dose (Figure 5).

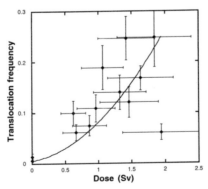

FIGURE 5. Translocation frequencies in Japanese A-bomb survivors measured using FISH with WCPs for chromosomes 1, 2, and 4. The curve is a dose-response curve measured for peripheral blood lymphocytes irradiated in vitro using FISH (Data from Lucas et al., 1992).

Some of this variation is probably due to errors in the estimation of the physical dose. However, part of this variation also may be due to individual variation in radiosensitivity. Thus, additional effort should be given to assessment of the extent of individual variability in radiosensitivity.

CANCER SPECIFIC ABERRATIONS

Cancer initiation is one important consequence of exposure to radiation. Thus, identification of specific genetic aberrations that contribute to carcinogenesis or to early stages of cancer progression may enable development of genetic assays for these events. One approach to the identification of such aberrations is to locate those that occur early in cancer progression. Aberrations that may be especially important include those that contribute to tumor suppressor gene inactivation or to the development of genetic instability (e.g., through inactivation or alteration of DNA repair genes such as MSH2 [Shibata et al., 1994] and MLH1 [Bronner et al., 1994] and cell cycle checkpoint genes such as p53 [Livingstone et al., 1992]). These may range from inactivating single base mutations to large deletions. Some of the involved genes are known but others remain to be discovered. Comparative genomic hybridization (CGH) is well suited to the localization of deletions or other changes in gene copy number that may contribute to gene inactivation or activation since it allows entire tumor genomes to be screened for increases and decreases in relative DNA sequence copy number in a single analysis (Kallioniemi et al., 1992). In CGH, differentially labeled samples of DNA from a tumor genome and a normal genome plus unlabeled cot-1 DNA are hybridized together to normal metaphase spreads (Kallioniemi et al., 1993). The rate of hybridization of the two labeled DNA samples to the target metaphase chromosomes is approximately proportional to their relative concentrations at each site in the genome. Differential labeling is usually accomplished by attaching fluorochromes that excite and fluoresce at different wavelengths to the two DNA samples so that the ratio of fluorescence from the two fluorochromes can be measured along each target chromosome using digital imaging microscopy as an indication of the relative DNA sequence copy number. Thus, CGH is well suited to the detection of DNA sequence losses that may contribute to tumor suppressor gene inactivation and gains that may contribute to oncogene activation. Of course, CGH detects only a subset of all possible aberrations and specifically does not respond to point mutations and balanced structural aberrations such as translocations and inversions.

CGH has been applied in studies of cancers of the breast (Kallioniemi et al., 1994), melanoma (Speicher et al., 1994), lung (Ried et al., 1993; Levin et al., 1994) and ovary (Sakamoto et al., personal communication). The number of aberrations detected using CGH is remarkable. It is common to find 10-15 different aberrations per tumor with more than half being whole chromosome or segmental reductions in relative copy number that may contribute to gene inactivation. However, this varies among tumor types. Some of the aberrations found commonly include increased relative DNA sequence number at the sites of known oncogenes such as Her-2/neu and CMYC and reduced relative DNA sequence copy number at sites of known tumor suppressor genes such as p53 (Kallioniemi et al., 1994). However, many of the DNA sequence copy number abnormalities involve regions of the genome not known to harbor cancer genes, suggesting that other cancer genes remain to be discovered. These studies illustrate the importance of CGH in identification of novel cancer genes and suggest several areas of research that may contribute to biomarker development including identification of aberrations that (a) consistently occur early in cancer progression, (b) correlate with genetic heterogeneity, or (c) occur consistently in cancers induced by specific DNA damaging agents such as radiation.

Early Genetic Aberrations

Definition of regions of common copy number abnormality that occur early in tumor progression may guide positional cloning of novel cancer genes. This may be accomplished by comparing aberrations found in low- and high-stage tumors. This has been difficult using CGH because current procedures require microgram quantities of DNA and thus are not well suited to the analysis of small lesions found in low-stage tumors. This may change in the future since some success has been achieved using PCR amplified genomic DNA for CGH analysis. At present, however, a better approach may be to shift to fluorescence *in situ* hybridization (FISH) with probes to regions of the genome suggested as being abnormal by CGH in late stage tumors. Once the involved genes are cloned, molecular or immunochemical bioassays for the frequency of cells carrying abnormal genes or gene products may predict development of the cancers initiated by these aberrations, and analysis of individuals carrying inherited genetic aberrations in these genes may identify persons that are particularly susceptible to development of specific cancers.

Aberrations Associated with Heterogeneity

The number of DNA sequence copy number aberrations detected by CGH and the degree of cell-to-cell variation measured using FISH vary substantially between tumors that are otherwise identical. We believe that the heterogeneity revealed by these techniques is an indication of the degree of genetic instability in the tumors (or alternately in the ability to remove genetically damaged cells). If true, definition of copy number aberrations that correlate with these endpoints may facilitate positional cloning of the involved genes. Assays for aberrations in these genes (p53, MSH2, and MHL1 are prototypic examples) may be used to detect increased risk of developing cancers that are initiated or that progress as a result of genetic instability.

Radiation Induced Genetic Aberrations

Identification of genetic aberrations specifically induced by ionizing radiation would facilitate discrimination between radiation-induced tumors and tumors that arise from lifetime exposure to other environmental agents. Whether such aberrations exist is unknown. However, the description of a mutation at codon 249 in p53 in lung cancers in miners exposed to radon (Taylor et al., 1994) but not in other lung cancers suggests the possibility that radiation specific abnormalities do exist. The difficulty is screening for presently unknown aberrations that may be radiation-specific. CGH may be a useful tool in this search since ionizing radiation often produces microdeletions that may be detectable by CGH as segmental changes in gene copy number. Identification of consistent copy number aberrations in human tumors in highly irradiated populations but not in matched tumors from non-irradiated populations would suggest the existence of radiation specific aberrations. Alternately, since CGH can be applied to mice as well as to humans, the existence of radiation-specific genetic changes in cancer may be assessed quickly through analysis of archived radiation-induced murine tumors.

CONCLUSIONS

Molecular cytogenetics techniques are powerful tools for the assessment of radiation-induced genetic damage in humans. FISH with WCPs facilitates measurement of the frequency of stable chromosome aberrations as an indication of level of exposure. However, scoring sufficient cells to enable accurate dose estimation is still a formidable task and can be accomplished only with appropriate automation. In addition, the biology of the hemopoietic system and its response to ionizing radiation is still poorly understood and requires further study using modern techniques. CGH is a powerful tool for discovery of novel genetic aberrations that may be cancer specific and/or radiation specific. Specific locus assays may be developed to detect mutations in genes discovered using CGH.

ACKNOWLEDGEMENTS

The authors appreciate pre-publication information from Dr. James Tucker regarding background translocation frequencies as well as numerous helpful discussions with Drs. Maria Pallavicini and Daniel Pinkel regarding the application of FISH to radiation dosimetry. Preparation of this manuscript was supported by the U.S. Department of Energy Contract DE-AC-03-76SF00098, NIH Interagency Agreement Y01-CP-10561, NIH grant P01-CA59431 and Imagenetics.

REFERENCES

Akiyama, M., Y. Kusunoki, S. Umeki, Y. Hirai, N. Nakamura, and S. Kyoizumi. 1992. Evaluation of four somatic mutations assays as biological dosimeters in humans. Pp. 172-176 in: Radiation Research: A Twentieth-Century Perspective, W. Dewey, et al., eds. Academic Press: San Diego.

Albertini, R. J., J. A. Nicklas, J. C. Fuscoe, T. R. Skopek, R. F. Branda, and J. P. O'Neill. 1993. In vivo mutations in human blood cells: biomarkers for molecular epidemiology. Environmental Health Perspectives 99:135-141.

Bender, M. A., A. A. Awa, A. L. Brooks, H. J. Evans, P. G. Groer, L. G. Littlefield, et al. 1988. Current status of cytogenetic procedures to detect and quantify previous exposures to radiation. Mutation Research 196:103-159.

Bronner, C. E., S. M. Baker, P. T. Morrison, G. Warren, L. G. Smith, M. K. Lescoe, et al. 1994. Mutation in the DNA mismatch repair gene homologue hMLH1 is associated with hereditary non-polyposis colon cancer. Nature 368:258-261.

Collins, C., W. L. Kuo, R. Segraves, J. Fuscoe, D. Pinkel, and J. W. Gray. 1991. Construction and characterization of plasmid libraries enriched in sequences from single human chromosomes. Genomics 11:997-1006.

Cowles, S. R., S. P. Tsai, P. J. Snyder, and C. E. Ross. 1994. Mortality morbidity and haematological results from a cohort of long-term workers involved in 13-butadiene monomer production. Occupational and Environmental Medicine 51:323-329.

Evans, J. W., J. A. Chang, A. J. Giaccia, D. Pinkel, and J. M. Brown. 1991. The use of fluorescence *in situ* hybridization combined with premature chromosome condensation for the identification of chromosome damage. British Journal of Cancer 63:517-521.

Grist, S. A., M. McCarron, A. Kutlaca, D. R. Turner, and A. A. Morley. 1992. *In vivo* human somatic mutation: frequency and spectrum with age. Mutation Research 266:189-196.

Jensen, R. H., S. G. Grant, R. G. Langlois, and W. L. Bigbee. 1991. Somatic cell genotoxicity at the glycophorin A locus in humans. Progress in Clinical and Biological Research 372:329-339.

Kallioniemi, A., O. P. Kallioniemi, D. Sudar, D. Rutovitz, J. W. Gray, F. Waldman, et al. 1992. Comparative genomic hybridization for molecular cytogenetic analysis of solid tumors. Science 258:818-21.

Kallioniemi, O. P., A. Kallioniemi, D. Sudar, D. Rutovitz, J. W. Gray, F. Waldman, et al. 1993. Comparative genomic hybridization: a rapid new method for detecting and mapping DNA amplification in tumors. Seminars of Cancer Biology 4:41-46.

Kallioniemi, O. P., A. Kallioniemi, J. Piper, L. Chen, H. Smith, J. W. Gray, et al. 1994. Chromosome mapping of amplified sequences in breast cancer by comparative genomic hybridization. Proceedings of the National Academy of Sciences (USA) 91:2156-2160.

Knehr, S., H. Zitzelsberger, H. Braselmann, and M. Bauchinger. 1994. Analysis for DNA-proportional distribution of radiation-induced chromosome aberrations in various triple combinations of human chromosomes using fluorescence in situ hybridization. International Journal of Radiation Biology 65:683-690.

Kyoizumi, S., S. Umeki, M. Akiyama, Y. Hirai, Y. Kusunoki, N. Nakamura, et al. 1992. Frequency of mutant T lymphocytes defective in the expression of the T-cell antigen receptor gene among radiation-exposed people. Mutation Research 265:173-180.

Langlois, R. G., M. Akiyama, Y. Kusunoki, B. R. DuPont, D. H. Moore, W. L. Bigbee, et al. 1993. Analysis of somatic cell mutations at the glycophorin A locus in atomic bomb survivors: a comparative study of assay methods. Radiation Research 136:111-117.

Levin, N., P. Brzoska, M. Warnock, J. W. Gray, and M. Christman. 1994. Identification of novel genetic alterations in small cell lung tumors. Cancer Research (in press).

Littlefield, L. G. and D. D. Lushbaugh. 1990. Cytogenetic dosimetry for radiation accidents — the good the bad and the ugly. Pp. 461-478 in: The Medical Basis for Radiation Accident Preparedness, R. C. Ricks and S. A. Fry, eds. Elsevier Science Publishing.

Littlefield, L. G., R. A. Kleinerman, A. M. Sayer, R. Tarone, and J. D. Boice. 1991. Chromosome aberrations in lymphocytes-biomonitors of radiation exposure. Pp. 387-397 in: Progress in Clinical and Biological Research. Volume 342. New Horizons in Biological Dosimetry, B. L. Gledhill and F. Mauro, eds. Wiley-Liss, New York.

Livingstone, L. R., A. White, J. Sprouse, E. Livanos, T. Jacks, and T.D. Tlsty. 1992. Altered cell cycle arrest and gene amplification potential accompany loss of wild-type p53. 70:923-935.

Lucas, J. N., A. Awa, T. Straume, M. Poggensee, Y. Kodama, M. Nakano, et al. 1992. Rapid translocation frequency analysis in humans decades after exposure to ionizing radiation. International Journal of Radiation Biology 62:53-63.

Matsuoka, A., J. D. Tucker, M. Hayashi, N. Yamazaki, and T. Sofuni. 1994. Chromosome painting analysis of X-ray-induced aberrations in human lymphocytes in vitro. Mutagenesis 9:151-155.

Motykiewicz, G., J. Michalska, J. Pendzich, F. Perera, and M. Chorazy. 1992. A cytogenetic study of men environmentally and occupationally exposed to airborne pollutants. Mutation Research 280:253-259.

Nakano, M., E. Nakashima, D. J. Pawel, Y. Kodama, and A. Awa. 1993. Frequency of reciprocal translocations and dicentrics induced in human blood lymphocytes by X-irradiation as determined by fluorescence in situ hybridization. International Journal of Radiation Biology 64:565-569.

Natarajan, A. T., R. C. Vyas, F. Darroudi, and S. Vermeulen. 1992. Frequencies of X-ray-induced chromosome translocations in human peripheral lymphocytes as detected by in situ hybridization using chromosome-specific DNA libraries. International Journal of Radiation Biology 61:199-203.

Perera, F., D. Tang, J. O'Neill, W. Bigbee, R. Albertini, R. Santella, et al. 1993. HPRT and glycophorin A mutations in foundry workers:relationship to PAH exposure and to PAH-DNA adducts. Carcinogenesis 14:969-973.

Pinkel, D., J. Landegent, C. Collins, J. Fuscoe, R. Segraves, J. Lucas, et al. 1988. Fluorescence in situ hybridization with human chromosome-specific libraries: detection of trisomy 21 and translocations of chromosome 4. Proceedings of the National Academy of Science (USA) 85:9138-9142.

Piper, J., and E. Granum. 1989. On fully automatic feature measurement for banded chromosome classification. Cytometry 10:242-55.

Ramalho, A. T., A. C. Nascimento, L. G. Littlefield, A. T. Natarajan, and M. S. Sasaki. 1991. Frequency of chromosomal aberrations in a subject accidentally exposed to ^{137}Cs in the Goiania (Brazil) radiation accident: intercomparison among four laboratories. Mutation Research 252:157-160.

Ried, T., I. Petersen, H. Holtgreve-Grez, M.R. Speicher, E. Schrock, S. du Manoir, et al. 1993. Mapping of multiple DNA gains and losses in primary small cell lung carcinomas by comparative genomic hybridization. Cancer Research 54:1801-1806.

Schmid, E., H. Zitzelsberger, H. Braselmann, J. W. Gray, and M. Bauchinger. 1992. Radiation-induced chromosome aberrations analyzed by fluorescence *in situ* hybridization with a triple combination of composite whole chromosome-specific DNA probes. International Journal of Radiation Biology 62:673-678.

Shibata, D., M. A. Peinado, Y. Ionov, S. Malkhosyan, and M. Perucho. 1994. Genomic instability in repeated sequences is an early somatic event in colorectal tumorigenesis that persists after transformation. Nature Genetics 6:273-281.

Speicher, M. R., G. Prescher, S. du Manoir, A. Jauch, B. Horsthemke, N. Bornfeld, et al. 1994. Chromosomal gains and losses in uveal melanomas detected by comparative genomic hybridization. Cancer Research 54:3817-3823.

Taylor, J. A., M. A. Watson, T. R. Devereux, R. Y. Michels, G. Saccomanno, and M. Anderson. 1994. P53 mutation hotspot in radon-associated lung cancer. Lancet 343:86-87.

Trask, B. and D. Pinkel. 1990. Fluorescence *in situ* hybridization with DNA probes. Methods in Cell Biology 33:383-400.

Tucker, J.D., D. A. Lee, M. J. Ramsey, J. Briner, L. Olsen, and D.H. Moore. 1994. On the frequency of chromosome exchanges in a control population measured by chromosome painting. Mutation Research (In press).

Weier, H. U., J. N. Lucas, M. Poggensee, R. Segraves, D. Pinkel, and J. W. Gray. 1991. Two-color hybridization with high complexity chromosome-specific probes and a degenerate alpha satellite probe DNA allows unambiguous discrimination between symmetrical and asymmetrical translocations. Chromosoma 100:371-376.

Wogan, G.N. 1992. Molecular epidemiology in cancer risk assessment and prevention: recent progress and avenues for future research. Environmental Health Perspectives 98:167-178.

Wolff, S., V. Afzal, and G. Olivieri. 1990. Inducible repair of cytogenetic damage to human lymphocytes: adaptation to low-level exposures to DNA-damaging agents. Progress in Clinical and Biological Research 340:397-405.

Wolff, S., R. Jostes, F. T. Cross, T. E. Hui, V. Afzal, and J. K. Wiencke. 1991. Adaptive response of human lymphocytes for the repair of radon-induced chromosomal damage. Mutation Research 250:299-306.

Wolff, S., V. Afzal, R. F. Jostes, and J. K. Wiencke. 1993. Indications of repair of radon-induced chromosome damage in human lymphocytes: an adaptive response induced by low doses of X-rays. Environmental Health Perspectives 101:73-77.

Biomarkers and Occupational Health: Progress and Perspectives. 1995. Pp. 215-225

Biomarkers to Detect Radiation Exposures

Karl T. Kelsey, John K. Wiencke, and Howard L. Liber

Recent developments in genetics and molecular biology now allow detailed characterization of genetic alterations which are occurring in normal human tissues as well as in cancer cells. Such information will be useful in improving assessments of the cancer risk associated with occupational exposures. Although many potential applications exist, the following discussion will focus on the specific area of worker health monitoring.

The overall goal of worker health monitoring is to identify individuals who exhibit either early signs of disease or a genetic predisposition to develop disease. It is, therefore, important at the outset to emphasize that worker health monitoring is not a form of primary prevention. Our discussion is focused on the development and use of biomarkers that will signal the need for intervention to prevent disease and to improve the conditions that lead to the identification of the health problem. Thus, all worker health monitoring, including that specifically employing molecular markers, should be performed only as an adjunct to ongoing primary prevention efforts.

It also seems appropriate to address the question of why it is important to continue to develop and improve biomarkers. Chief among the reasons is the fact that the existing biomarkers are largely non-specific. Ionizing radiation and many hazardous chemicals are genotoxic; therefore, exposure to workers can induce genetic damage. Assays abound that can detect nonspecific genetic damage. There are several that are useful for detecting worker exposure. These assays primarily use peripheral blood lymphocytes as cellular surrogates for target tissues and can be used to detect many forms of DNA damage. Some of the assays that have been developed utilize cytogenetic techniques including the induction of sister chromatid exchanges, micronuclei, or classical chromosome aberrations. Other assays measure the induction of specific somatic gene mutations as an endpoint of *in vivo* genotoxic exposures. Two of the gene loci examined include hypoxanthine phosphoribosyl transferase (HPRT) and the HLA locus. Both of these mutation assays examine peripheral blood lymphocytes to determine the induction of mutations (Albertini, 1985; Albertini et al., 1982, 1985; McCarron et al. 1989; Morley et al., 1982, 1983, 1985; Strauss and Albertini, 1977, 1979). Alternatively, mutations of the glycophorin locus can also be measured in red blood cells, and this technique has been applied to radiation-exposed populations (Jensen et al., 1987; Kyoizumi et al., 1989; Langlois et al., 1987; Nakamura et al., 1991). Newer assays involving other loci are being developed, including an antibody-based assay for mutations in T-cell receptor genes (Kyoizumi et al., 1990, 1992).

While all of these assays can detect the genetic consequences of exposure to genotoxic agents, they are relatively non-specific in that they reveal nothing about the identity of the genotoxic agent. Therefore, when workers are exposed to multiple contaminants, as is commonly the case in hazardous waste site remediation and other occupational settings, the presence of these biomarkers will indicate that there has been exposure to genotoxic chemicals but will not provide information on which chemical is associated with the observed effects. As a result, a positive response in the biomarker assay would necessarily prompt efforts to reduce all exposures. In complex workplaces, this approach may not be possible. Thus, it is of particular interest to develop molecular biomarkers that will be able to distinguish exposure to ionizing radiation from exposure to specific toxic chemicals.

Current Approaches to Biomarker Development

The current strategies for the development of biomarkers have followed a common pathway. Research is now being aimed at the determination of signature genomic lesions from within the spectrum of radiation-induced abnormalities. Radiation induces a wide variety of DNA lesions (Teoule, 1987), several of which may be converted into stable genetic alterations. Detection of a unique alteration caused solely by radiation presents a difficult problem.

The development of biomarkers for monitoring workers presents additional problems, since the signature lesion should ideally be quantitatively relatable to levels of exposure or at least qualitatively relatable to a health effect. Also, a validated biomarker should provide useful information beyond that obtained by ambient monitoring. Thus, biomarkers under development must meet stringent criteria if they are to be useful as effective health monitors in the workplace.

Among the most promising current strategies are those that use mutational lesions as candidates for development of radiation-specific biomarkers. Mutational events are an integral part of the carcinogenic process and are readily produced by radiation exposure (Graf and Chaisin, 1982; Grosovsky et al., 1986; Stankowski and Hsie, 1986; Vrieling et al., 1985). It is believed that multiple mutational events are necessary for a single cell to clonally develop into a malignancy (Vogelstein et al., 1988). Detection of mutational events may also occur early in the process of exposure-induced malignancy and, therefore, may be an excellent candidate for monitoring. Certain mutations in specific malignancies have been shown to be related to clinical outcome in many diseases. Therefore, the detection of mutations caused specifically by radiation may have some relationship to the health of the exposed worker.

A relevant illustration is occupational and chemotherapy-related acute non-leukocytic leukemias. Of all the human cancers, leukemias have been the subject of the most extensive genetic analyses of the accompanying genomic alterations. Association between occupational exposures and specific alterations in diseased tissues have been observed for cytogenetic aberrations and recently for point mutations in members of the *ras* family of oncogenes (Taylor et al., 1992). Studies conducted in the 1970s in rodents and humans suggest that tumors could show a chromosomal pattern specific for the cancer-inducing agent. As cytogenetic analyses of bone-marrow specimens have become a routine part of the clinical approach to leukemia diagnosis and therapy, an extensive database has been compiled. This information has ultimately led to the recognition that histological subtypes of leukemias are associated with specific chromosomal abnormalities (LeBeau and Larson, 1991). Several studies have estimated the prevalence of occupational chemical exposures in patients stratified according to the presence of cytogenetic abnormalities. In these studies, patients with acute non-lymphocytic leukemia (ANLL) who reported occupational exposure to chemical solvents were more likely to have one of four clonal chromosomal abnormalities (Crane et al., 1989; Golomb et al., 1982; Mitelman et al., 1978, 1981; Narod and Dube, 1989).

Other studies have shown that abnormalities of chromosome 5 or 7 account for the majority of clonal aberrations in myelodysplastic syndromes that often evolve into ANLL (Geddes et al., 1990). The variation in the location of break-points in the deletions of chromosomes 5q and 7q in myelodysplasia and acute leukemias related to prior anti-cancer therapies indicate that the inactivation of genetic loci, rather than the specific activation of proto-oncogenes, may be involved in these cancers (Pedersen-Bjergaard and Philip, 1987; Rowley and LeBeau, 1989). Further, significant geographic heterogeneity of neoplasm-associated chromosomal abnormalities in hematologic and solid tumors has been noted (Johansson et al., 1991). Thus, specific genetic patterns that have been identified through the use of large databases can be useful in generating hypotheses about the prevalence of exposures associated with cancers in different populations.

Mutational Spectra

The development of biomarkers specific for radiation effects has currently concentrated on the use of mutational spectra. A mutational spectrum can be defined as a set of all mutations found in a defined DNA sequence in a population of cells (Thilly et al., 1989). These mutations can include base-pair substitutions, insertions, deletions, and larger chromosomal changes, including large-scale deletions and insertions and rearrangements. Historically, mutations were classified in two groups: intragenic mutations and those that are intergenic or the result of chromosomal rearrangements and losses. These two classes of mutations are not usually mutually exclusive but blend into one another as the size of the intergenic mutation decreases.

Currently, a large research effort has been mounted to examine the spectrum of mutations induced in loci that are thought to be causally involved in the generation of malignant diseases. This research has as its goal the definition of a spectrum that is specific for exposure. In this way, retrospective knowledge of causal exposures can be gleaned from examination of diseased tissue.

More germane to the current topic is the effort to use "reporter genes" in the analysis of mutational spectra. In studies of worker exposure to carcinogens, this strategy for assessing the ability of compounds to induce specific genetic alterations is rapidly emerging. This approach uses reporter genes as the DNA target and assumes that these genes, which are not in the causal pathway for disease, undergo genetic alterations that "report" the ability of the compound tested to induce damage in the genes that are involved in cancer. A common gene employed for mutational spectra analysis is the HPRT gene in lymphoblastoid cell lines and peripheral blood lymphocytes.

Currently, significant effort is under way to generate a database on these loci for future studies of worker exposure. After sufficient numbers of mutants have been analyzed, it may be possible to use the spectral data to generate a list of agents that could have induced the mutants recovered from any exposed individual under study. To date, approximately 1,200 HPRT mutants have been analyzed from both lymphoblastoid cells and from peripheral blood lymphocytes isolated from *in vivo* situations.

Our current research involves the detection of radiation signature lesions using HPRT mutants initially generated *in vitro*. The selection of mutants is based upon 6–thioguanine incorporation. Since the HPRT gene is well characterized, the precise nature of the lesion can be identified. Radiation-induced mutants have been sequenced and the lesions identified have been compared to the database of spontaneously-arising and chemically-induced mutants. Figure 1 depicts the three distinct mutants induced by radiation that have been isolated. These are thought to be radiation-specific lesions.

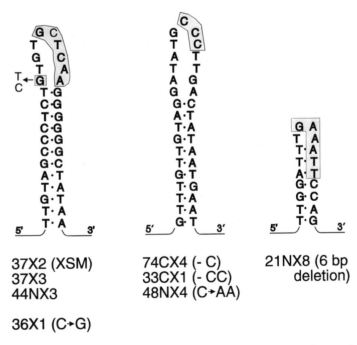

37X2 (XSM)   74CX4 (- C)   21NX8 (6 bp
37X3   33CX1 (- CC)   deletion)
44NX3   48NX4 (C→AA)

36X1 (C→G)

FIGURE 1. Putative secondary structure leading to three radiation-induced HPRT mutants. These mutants are candidates for radiation-specific lesions.

These lesions can serve as the basis for the development of techniques to identify them in a background of normal cells. Research is now under way to develop polymerase chain reaction-based methods to identify these signature mutants in the background of normal cells. In this way, an assay for the early identification of radiation-specific mutations may be applied as a biomarker of radiation exposure and effect in workers. Current work has involved reconstruction experiments that have demonstrated that we can reliably detect one mutant in a background of 10^4 cells (Figure 2). This level of sensitivity is on the threshold of that needed for field testing of these methods.

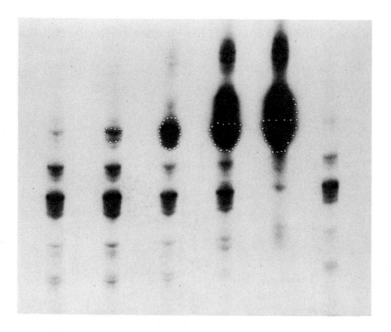

FIGURE 2. ^{32}P-Labeled PCR products from oligo-specific amplification of one radiation-specific lesion. The first lane is all wild-type cells. The second lane is all mutant cells. Lanes 3-6 are serial dilutions of mutant cells in wild-type background (1 in 10, 1 in 10^2, 1 in 10^3 and 1 in 10^4).

Future Needs

The level of sensitivity of current methods has been reported to be sufficiently high to detect one specific mutant in a background of

10^5 cells. At this level, it is practical to believe that application of this method in a working population is possible. However, major questions remain to be addressed. The method is likely to be reliable, reproducible, and inexpensive. It is, therefore, likely to be acceptable for potential use as a routine monitoring tool. However, it is unclear if these signature lesions occur in the peripheral blood lymphocytes with any frequency. Thus, significant work remains to demonstrate that the use of assays such as the ones discussed above, will be at all sensitive. It is likely then that methods will be used in DNA derived from lymphocytes and the kinetics of turnover lymphocytes have certain implications for the detection of radiation effects. Further, it also remains to be shown that this lesion is truly specific. Finally, it is important to show that the presence of the lesion is related to health outcomes and that, in addition, it provides useful information beyond that obtained by ambient monitoring.

CONCLUSIONS

Current strategies for the development of radiation-specific biomarkers are progressing rapidly, especially those employing the concept of mutational spectral analysis. Signature lesions have been identified and polymerase chain reaction methods are showing great promise for use in detecting these lesions in a background of normal cells. Significant research questions must be answered, however, before these biomarkers can find practical application in the workplace. At the same time, the path toward development of these biomarkers seems clear and the research is progressing at a rapid pace.

ACKNOWLEDGEMENTS

Supported by grants CA 56420 from the NCI, ES-00001 from the NIEHS and K01 OH 00110 from the NIOSH.

REFERENCES

Albertini, R. J. 1985. Somatic gene mutations *in vivo* as indicated by the 6-thioguanine–resistant T-lymphocytes in human blood. Mutation Research 150:411.

Albertini, R. J., K. L. Castle, and W. R. Borcherding. 1982. T-cell cloning to detect the mutant 6-thioguanine-resistant lymphocytes present in human peripheral blood. Proceedings of the National Academy of Sciences (USA) 79:6617.

Albertini, R. J., L. M. Sullivan, J. K. Berman, et al. 1988. Mutagenicity monitoring in humans by autoradiographic assay for mutant T lymphocytes. Mutation Research 204:481.

Crane, M. M., M. J. Keating, J. M. Trulillo, et al. 1989. Environmental exposures in cytogenetically-defined subsets of acute non-lymphocytic leukemia. Journal of the American Medical Association 262:634-639.

Geddes, A. D., D. T. Bowen, and A. Jacobs. 1990. Clonal karyotype abnormalities and clinical progress in the myelodysplastic syndrome. British Journal of Haematology 76:194-202.

Golomb, H. M., G. Alimena, J. D. Rowley, J. W. Vardman, J. R. Testa, and C. Sovik. 1982. Correlation of occupation and karyotype in adults with acute non-lymphocytic leukemia. Blood 60:404-411.

Graf, L. H., and L. A. Chaisin. 1982. Direct demonstration of genetic alteration at the dihydrofolate reductase locus after γ-irradiation. Molecular and Cellular Biology 2:93.

Grosovsky, A. J., E. A. Drubetsky, P. J. deJong, et al. 1986. Southern analysis of genomic alterations in γ-ray induced HPRT-hamster cell mutants. Genetics 113:405.

Jensen, R. H., W. L. Bigbee, R. G. Langlois. 1987. *In vivo* somatic mutations in the Glycophorin A locus of human erythroid cells. Pp. 139-148 in M. Moore, D. Tindall, F. Demarini and F. DeSerres, eds. Mammalian Cell Mutagenesis, Banbury Reports 28, Cold Spring Harbor Laboratory, Cold Spring Harbor.

Johansson, B. F., F. Martens, and F. Mitelman. 1991. Geographic heterogeneity of neoplasia-associated chromosome aberrations. Genes, Chromosomes and Cancer 3:1-7.

Kyoizumi, S., N. Nakamura, M. Hakoda, et al.: 1989. Detection of somatic mutations at the glycophorin A locus in erythrocytes of atomic bomb survivors using a single beam flow sorter. Cancer Research 49:581.

Kyoizumi, S., M. Akiyama, Y. Hirai, et al. 1990. Spontaneous loss and alteration of antigen receptor expression in mature CD4[+] T cells. Journal of Experimental Medicine 171:1981.

Kyoizumi, S., S. Umeki, M. Akiyama, et al. 1992. Frequency of mutant T lymphocytes defective in the expression of the T-cell antigen receptor gene among radiation-exposed people. Mutation Research 265:173.

Langlois, R. G., W. L. Bigbee, S. Kyoizumi, et al. 1987. Evidence for increased somatic cell mutations at the glycophorin A locus in atomic bomb survivors. Science 236:445.

LeBeau, M. M., and R. A. Larson. 1991. Cytogenetics and neoplasia. Pp. 638-655 in R. Hoffman, E. J. Benz, Jr., S. J. Shattil, B. Furie, H. J. Cohen, eds. Hematology, Basic Principles and Practice, New York: Churchill Livingstone.

Liber, H. L., K. M. Call, and J. B. Little. 1987. Molecular and biochemical analyses of spontaneous and X-ray-induced mutants in human lymphoblastoid cells. Mutation Research 178:143.

McCarron, M. A., A. Kutlaca, and A. A. Morley. 1989. The HLA-A mutation assay: Improved technique and normal results. Mutation Research 225:189-193.

Mitelman, F., L. Brandt, and P. G. Nilsson. 1978. Relation among occupational exposure to potential mutagenic/carcinogenic agents, clinical findings, and bone marrow chromosomes in acute non-lymphocytic leukemia. Blood 52:1229.

Mitelman, F., P. G. Nilsson, L. Brandt, G. Alimena, R. Gastaldi, and B. Dallapiccola. 1981. Chromosome pattern, occupation, and clinical features in patients with acute non-lymphocytic leukemia. Candian Journal of Genetics and Cytogenetics 4:197-214.

Morley, A. A., S. Cox, and D. Wigmore. 1982. Enumeration of thioguanine-resistant lymphocytes using autoradiography. Mutation Research 95:363.

Morley, A. A., K. J. Trainor, R. Seshadri, et al. 1983. Measurement of in vivo mutations in human lymphocytes. Nature (London) 302:155.

Morley, A. A., K. J. Trainor, J. L. Dempsey et al. 1985. Methods for study of mutagenesis in human lymphocytes. Mutation Research 147:363.

Nakamura, N., S. Umeki, Y. Hirai, et al. 1991. Evaluation of radiation-exposed people including atomic bomb survivors. Progress in Clinical Biological Research 372:341.

Narod, S. A., and I. D. Dube. 1989. Occupational history and involvement of chromosomes 5 and 7 in acute non-lymphocytic leukemia. Canadian Journal of Genetics and Cytogenetics 38:261-269.

Pedersen-Bjergaard, J., and P. Philip 1987. Cytogenetic characteristics of therapy-related acute non-lymphocytic leukemia, pre-leukemia and acute myeloproliferative syndrome: correlation with clinical data for 61 consecutive cases. British Journal of Haematology 66:199-207.

Rowley, J. D., and M. M. LeBeau. 1989. Cytogenetic and molecular analysis of therapy-related leukemia. Annals of the New York Academy of Sciences 567:130-140.

Stankowski, L. F., Jr., and A. W. Hsie. 1986. Quantitative and molecular analyses of radiation-induced mutation in AS52 cells. Radiation Research 105:37.

Strauss, G. H., and R. J. Albertini. 1977. 6-Thioguanine resistant lymphocytes in human peripheral blood. Pp. 327-334 in: Progress in Genetic Toxicology, D. Scott, B. A. Bridges, and F. H. Sobels, eds. Amsterdam, Elsevier/North-Holland Biomedical Press.

Strauss, G. H., R. J. Albertini. 1979. Enumeration of 6-thioguanine-resistant peripheral blood lymphocytes in man as a potential test for somatic cell mutations arising in vivo. Mutation Research 61:353.

Taylor, J. A., D. P. Sandler, C. D. Bloomfield, et al. 1992. Ras oncogene activation and occupational exposures in acute myeloid leukemia. Journal of the National Cancer Institute 84:1626-1632.

Teoule, R. 1987. Radiation-induced DNA damage and its repair. International Journal of Radiation Biology 51:573.

Thilly, W. G., V. F. Liu, B. J. Brown, N. F. CaAello, A. G. Kat, and P. Keohavong. 1989. Direct measurement of mutational spectra in humans. Genome 31:590-593.

Vogelstein, B., E. R. Fearon, S. R. Hamilton, et al. 1988. Genetic alterations during colorectal-tumor development. New England Journal of Medicine 319:525.

Vrieling, H., J.W. Simons, and F. Arwert. 1985. Mutations induced by X-rays at the HPRT locus in cultured Chinese hamster cells are mostly large deletions. Mutation Research 144:277.

Biomarkers and Occupational Health: Progress and Perspectives. 1995. Pp. 226-237

Dioxin Congeners Distribution in Biological Samples as Biomarkers for Exposure

Venkateswara Rao and Alan Unger

The isomeric compositions of polychlorinated dibenzo-p-dioxin and dibenzofuran congeners (PCDDs) vary with sources such as incinerators, industrial effluents, chemicals containing PCDDs as contaminants, and food samples (Safe, 1990). Table 1 illustrates the isomeric composition of PCDDs, expressed as percent of the total mass, for (a) soil samples from a wood preservative site contaminated with dioxins (EPA, 1992a), (b) emissions from a hazardous waste incinerator (Rao et al., 1994a), and (c) dioxin-contaminated fish samples (EPA, 1992b). The PCDD distributions and their relative concentrations are characteristic for these sources. For example, octachlorodibenzo-p-dioxin (OCDD) is the most prevalent congener, accounting for up to 66% of the total mass in the soil samples from the wood preservative site, whereas OCDD is almost non-existent in the incinerator emissions and fish samples. Higher congeners of dibenzofuran comprise the bulk of the incinerator emissions and account for up to 88% of the total mass, whereas their levels are low in the soil (11%) and fish (16%) samples. High levels of more toxic congeners of PCDDs, particularly 2,3,7,8-tetrachlorodibenzo-p-dioxin (TCDD) (13%), and 2,3,7,8-tetrachlorodibenzofuran (TCDF) (19%) have been reported in fish samples, whereas their levels are below detection limits in the incinerator samples and in soil samples from a wood preservative site (EPA, 1992a).

Distribution of Dioxins in Biologic Samples

The relative distribution of PCDDs in biological organisms is a complex process governed by several extrinsic and intrinsic factors. Physicochemical properties of PCDDs determine their distribution in environmental media. Those properties, as well as differences in the physiological and biological processes characteristic of the exposed organisms, determine the overall residue levels in the tissues of those organisms.

For example, the PCDD composition in fish samples will be comparable to the sources of contamination in aquatic media, whereas levels of
PCDDs in biological samples from humans (e.g., human adipose tissue
and milk) are not so clearly related to levels in the source of exposure
(Safe, 1990). Analyses of adipose and milk samples have revealed elevated concentrations of the more toxic 2,3,7,8-substituted congeners,
although inactive OCDD is consistently the most prevalent congener
detected in the human samples.

TABLE 1. ISOMERIC COMPOSITION OF PCDD CONGENERS CHARACTERISTIC
OF THE ENVIRONMENTAL SOURCE

	Mean Percent of the Total PCDD Mass in the Sample		
Congeners	Wood Preservative[1] Site (n = 29)	Hazardous Waste[2] Incinerator (n = 5)	Fish[3] (n = 116)
2,3,7,8-TCDD	ND	0.95	13.1
1,2,3,7,8-PenCDD	ND	3.01	7.3
1,2,3,6,7,8-HexCDD	0.66	5.31	12.7
1,2,3,4,6,7,8-HepCDD	13.93	0.7	32.0
OCDD	65.69	0.09	ND
2,3,7,8-TCDF	ND	2.05	18.9
2,3,4,7,8-PenCDF	0.07	29.49	7.2
1,2,3,6,7,8-HexCDF	2.33	24.17	4.2
1,2,3,4,6,7,8-HepCDF	8.36	33.99	4.6
OCDF	8.96	0.23	ND

Based on [1]EPA (1992a); [2]Rao et al. (1994c); [3]EPA (1992b).

We examined the possible use of the PCDD composition in biological samples as a marker for the source type. The underlying assumption
was that the source-dependent differences in the PCDD profile would
be reflected in their composition in biological tissues. Although the isomeric composition in human tissue samples does not provide clues as to
the source(s) of exposure, it has been suggested that apart from occupational exposure category, dietary intake constitutes the largest source
of exposure to PCDDs in the general population (Beck et al., 1989).
Fish, dairy products, poultry, beef, and produce have been shown to

have widely varying isomeric composition of PCDDs (Beck et al., 1989; Goldman et al., 1989).

There are only a few reports on biomarkers for exposure to PCDDs (Neubert et al., 1993; Weber et al., 1992). Immunoassays of cell surface receptors from lymphocyte samples collected from humans volunteers with moderately increased body burdens of dioxins have indicated that $CD45R0+$, a cell surface receptor of the helper-induced (memory) T cells, exhibited slight association with high exposure (10-23 ppt TCDD) group (Neubert et al., 1993). Some investigators have suggested alkoxyresorufin metabolism as a nonspecific biomarker for exposure to chemicals that induce arylhydrocarbon (Ah) receptor activity (Lubet et al., 1990). In a laboratory study, male Sprague-Dawley rats exposed to a mixture of tetra-, penta-, hexa-, and hepta-congeners of dioxin showed a dose-dependent elevation in the plasma tryptophan levels. A 28-fold increase in ethoxyresorufin-o-deethylase (EROD) activity was also observed in the treated group. However, there was no correlation between plasma tryptophan levels, a biomarker for acute toxicity of dioxins, and EROD activity, a biomarker for Ah receptor-mediated enzyme induction (Weber et al., 1992).

As a part of an ongoing effort to study the effects of exposure to combinations of chemicals on human carcinogenesis, we have been attemping to characterize exposure to multiple congeners of PCDDs (Rao et al., 1989; Rao, 1991a,b; 1992). Our objective was to examine whether composition of PCDDs in the human biological samples could be used to: (a) identify exposed subgroups from a larger human population sample, and (b) examine tissue from the exposed group to determine environmental sources of PCDDs. Since isomer compositions in human samples (adipose, milk) are not indicative of the source type, we examined the possibility of comparing a model-derived Ah receptor-congener bound fraction, defined as f value, for various PCDDs to distinguish the source(s) of exposure. A competitive binding model had been developed earlier to estimate cellular-level doses of this congener when a person was exposed to a mixture of PCDDs (Rao and Unger, 1994b).

Use of Model Algorithms to Derive Cellular-Level Doses

To model the fraction of Ah receptor bound to TCDD (at concentration $[L]$) in the presence of a set of n competing congeners (at concentrations $[C_i]$, $i = 1, ..., n$), Equation (1) was derived as the generalized algorithm to model competitive binding to Ah receptor.

$$f = \frac{[RL]_n}{[RL]_n + \sum_{i=1}^{n} [RC_i]_n} = \frac{\frac{[L]_n}{K_L}}{\frac{[L]_n}{K_L} + \sum_{i=1}^{n} \frac{[C_i]_n}{K_i}} = \frac{1}{1 + \sum_{i=1}^{n} \frac{[C_i]_n}{[L]_n} \frac{K_L}{K_i}}$$

Where: f = Estimated ligand-receptor bound fraction; $[RL]$ = receptor-ligand bound concentration; $[RC_i]$ = receptor-competitor (i = 1...n); K_i = dissociation constant of competitor (i = 1....n); K_L = dissociation constant for ligand; $[L]$ = ligand concentration; $[C_i]$ = free competitor (i = 1...n) concentration.

By replacing the ratio K_L/K_C in the above equation by $ED_{50}(L)/ED_{50}(C)$ (ED50=dose at which 50% of the exposed organisms demonstrate a given biological effect), we obtain an expression that can be used to estimate the fraction of receptor that is bound to TCDD in a mixture of PCDDs with known concentrations. The ED_{50} values were obtained from competitive binding experiments using rat cytosolic Ah receptor preparations (Haake et al., 1987; Biegel et al., 1989).

In this model, differential binding affinities of PCDDs to the Ah receptor and the differences in the relative masses of PCDDs are the basic elements of the competitive binding algorithms. Based on a series of algorithms, a numerical index, f value, was derived for the individual congeners. These values represent the maximal likelihood estimate for the formation of a congener-receptor bound complex in the presence of other competing ligands. Thus, for a mixture of PCDDs in the adipose tissue survey data, a set of f values for individual congeners was derived.

Application of the Model Algorithm

To test the application of this model to distinguish the exposed population based on the source of exposures, two distinct exposure scenarios were considered. First, mean human adipose tissue residue data published for a sample population from the People's Republic of Vietnam, Germany, Japan, and the United States were used to generate f values and fractional cellular-level doses of PCDDs (Rappe et al., 1987). Second, adipose tissue data from a sample population from North and

South Vietnam were used to derive f values to distinguish background and exposed populations. Earlier reports have implicated the spraying of Agent Orange, a phenoxy herbicide contaminated with TCDD, with differences observed in the PCDDs profile in the sampled groups (Schecter et al., 1986). The rationale for the selection of these data sets was to examine the possibility of applying the f values as a marker for exposure to PCDDs.

Based on the serum half-life of 7.1 years for TCDD, a simple first-order elimination equation was used to derive serum concentrations of PCDDs as a fraction of their adipose tissue levels. Model algorithms were used to calculate f values to estimate TCDD toxicity equivalence (TEF) doses according to an EPA-recommended method (Stanley et al., 1986). To determine an approximate confidence interval on the fraction of bound receptors in a given mixture, it was assumed that the uncertainty in concentrations of the PCDDs can be lognormally modeled. A Monte Carlo simulation was used to estimate the resulting distribution of the fraction bound, from which an approximate confidence interval was obtained. Differences between the f values for congeners in population subgroups imply that these groups were exposed to different sources. The f values for individual congeners computed from the competitive binding algorithms were used to derive the fractional cellular-level doses.

Results and Discussion

Competitive binding model algorithms were used to estimate cellular-level doses of PCDDs based on the f values for congeners detected in the human adipose tissue samples for populations from the Republic of Vietnam, Germany, Japan, and the United States. Figure 1 illustrates the percent mean values for the EPA-recommended TEF doses for PCDDs (1A) and fractional cellular-level doses using the model algorithms (1B) for the adipose tissue residue data from 4 countries. Comparison of adipose tissue profile in the figure indicated that:

- Except for the TCDD fraction in the South Vietnamese sample, no congener fraction(s) could be assigned as unique in the adipose tissue data from other countries using the standard TEF method (Figure 1A).

- Percent mean f values and cellular-level doses of various congeners were distinct for various sampled populations. Use of the f values of PCDDs enhanced the possibility of associating adipose tissue residues with potential source(s) of PCDDs. For example, the model algorithms estimated the highest f value of 86.3% for TCDD which corresponds to 98.8% of the fractional cellular-level dose (Figure 1B). Exposure to TCDD-contaminated Agent Orange was implicated as the source of the elevated TCDD residue levels in the South Vietnamese adipose tissue samples (Rappe et al., 1987; Schecter et al., 1986).

- The f values from the adipose residue data from Germany, Japan, and U.S. samples displayed the highest value for 2,3,4,7,8-penta-chloro-dibenzofuran (penCDF). For example, the f values were 38.3%, 49.8%, and 28%, which corresponded to fractional cellular-level doses of 76.6%, 86.2%, and 50.8% for the German, Japanese, and U.S. samples, respectively. Although penCDF is a ubiquitous congener, its levels are particularly elevated in incinerator emissions (Table 1). Earlier studies have pointed out incinerators as the source for the presence of high levels of this congener in the milk samples from Germany (Beck et al., 1989).

- High f value for the inactive OCDD ($f = 4.53\%$) in the Japanese samples correlated with the high levels of this congener in the sampled population (Rappe et al., 1987). Earlier studies have implicated high levels of fish consumption by the Japanese as a possible source of the elevated levels of OCDD in their adipose tissue (Rappe et al., 1987) and milk samples (Schecter et al., 1989).

- The percent fractional distributions of PCDDs were generally comparable for the samples collected from the industrialized countries of Germany, Japan, and United States, and differed considerably from the Vietnamese data. The samples from industrialized countries showed consistently elevated levels of more toxic 2,3,7,8-substituted higher PCDDs. Among the industrialized countries, U.S. samples had the highest levels of TCDD ($f = 29.1\%$) with a high cellular-dose fraction (37.2%), which was followed by Germany ($f = 19.8\%$) and Japan ($f = 17.6\%$). This observation corroborates an earlier study reporting elevated levels of higher congeners in the milk specimens from industrialized countries compared with samples from a developing country (Schecter et al., 1989).

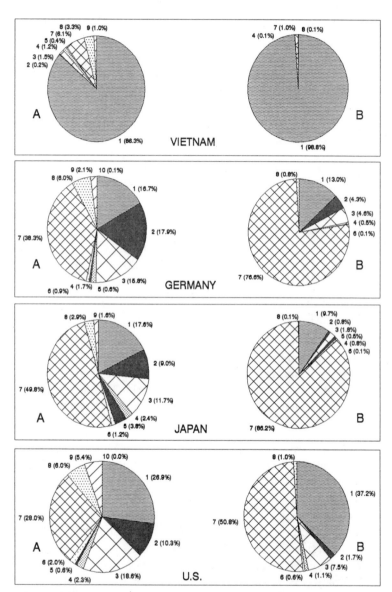

1 = 2,3,7,8-TCDD; 2 = 1,2,3,7,8-PenCDD; 3 = 1,2,3,6,7,8-HexCDD; 4 = 1,2,3,4,6,7,8-HepCDD;
5 = OCDD; 6 = 2,3,7,8-TCDF; 7 = 2,3,4,7,8-PenCDF; 8 = 1,2,3,4,7,8/1,2,3,6,7,8-HexCDF;
9 = 1,2,3,4,6,7,8-HepCDF; 10 = OCDF.

FIGURE 1. Comparison of mean toxicity equivalence factor-dose (TEF-dose) [A] and mean fractional cellular-level doses [B] of PCDD congeners in adipose tissue samples from four countries.

TCDD was the most prominent congener in the South Vietnamese samples ($f = 78.8\%$), which was also the highest value among all the samples selected for comparison. The highest f value for 1,2,3,7,8-penCDD ($f = 6.19\%$) was detected in the German samples and highest OCDD ($f = 4.53\%$) was found in the Japanese data. Vietnamese samples revealed the lowest f values for the 1,2,3,7,8-penCDD (0.09%), 1,2,3,6,7,8-hexCDD ($f = 0.81\%$), and penCDF ($f = 11\%$) compared with those for industrialized countries; 1,2,3,7,8-penCDD (Germany = 6.19%, Japan = 2.71%, U.S. = 3.49%), 1,2,3,6,7,8-hexCDD (Germany = 7.46%, Japan = 4.93%, U.S. = 8.44%), and penCDF (Germany = 50.99%, Japan = 55.23%, U.S. = 38.17%).

In addition, the adipose tissue residue survey of the Vietnamese population was used to test the applicability of f values to distinguish the exposed group, potentially exposed to TCDD-contaminated Agent Orange, with a comparable background population (Rappe et al., 1987; Schecter et al., 1986). Figure 2 illustrates the model-derived percent mean fractional tissue doses of PCDDs for the exposed and background groups. The mean adipose residue concentrations of PCDDs in the exposed and control groups were almost identical (Figure 2A). Although the TCDD concentration in the exposed group was 8-fold higher than the control, the difference is not clearly demonstrable due to very high levels of higher congeners. Since TCDD accounts for only 5% of the total mass in the exposed group and 0.8% in the control group, the combined profiles of PCDDs appear similar.

In comparison, the f values and fractional cellular-level doses of PCDDs were unique for the exposed and background groups (Figure 2B). In the exposed group, TCDD accounted for up to 62% of the total Ah receptor-bound fraction which corresponds to 92% of the fractional cellular-dose, whereas the f values for more ubiquitous congeners like OCDD and penCDF were 0.9 and 19.1%, respectively; the fractional cellular-level dose of penCDF was 5.1%. For the background group (Figure 2B), TCDD fraction represented only 9.6% of the total cellular-level dose. Other higher congener of dioxins such as 1,2,3,7,8-PenCDD ($f = 7.3\%$) and 1,2,3,6,7,8-HexCDD ($f = 4.6\%$) accounted for up to 15% of the cellular-level dose. However, penCDF ($f = 43\%$) was the most predominant congener in the background population accounting for up to 72% of the total fractional tissue dose. The high f value for penCDF in the background Vietnamese population corresponds fairly well with those of the German (76.6%), Japanese (86.2%), and U.S. (50.8%) populations (Figure 1). However, the overall mass of PCDDs in the Vietnamese samples are low compared with those from industrialized countries. Higher f values and fractional tissue doses of higher toxic congeners of PCDDs appear to be characteristic of the background population. Occupational or other site-specific exposure to TCDD-contaminated matrices is associated with increases in the f values and fractional cellular-level doses. Differences between exposed and background groups may be established when f values were used as the marker for exposure. These results are of a preliminary nature since there are very few well-conducted human tissue sample surveys of PCDDs or toxicokinetics studies with PCDD mixtures comparing administered dose and tissue-concentration relationships.

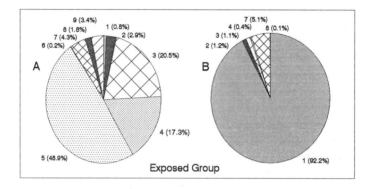

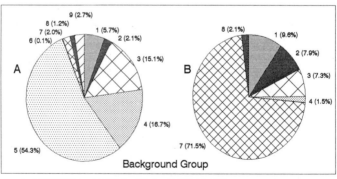

1 = 2,3,7,8-TCDD; 2 = 1,2,3,7,8-PenCDD; 3 = 1,2,3,6,7,8-HexCDD; 4 = 1,2,3,4,6,7,8-HepCDD;
5 = OCDD; 6 = 2,3,7,8-TCDF; 7 = 2,3,4,7,8-PenCDF 8 = 1,2,3,4,7,8/1,2,3,6,7,8-HexCDF;
9 = 1,2,3,4,6,7,8-HepCDF; 10 = OCDF.

FIGURE 2. Comparison of mean adipose tissue residue levels (ppt) [A] and model-derived mean fractional tissue doses of PCDDs [B] based on adipose tissue residue surveys of exposed and background groups.

ACKNOWLEDGEMENTS

The authors would like to acknowledge Dr. M. Owens, Mr. M. Rager, and Dr. E. Straker, of SAIC, for the financial support for the preparation and presentation of this manuscript.

REFERENCES

Beck, H., K. Eckart, W. Mathar, and R. Wittkowski. 1989. PCDD and PCDF body burden from food intake in the Federal Republic of Germany. Chemosphere 18:417-424.

Biegel, L., M. Harris, D. Davis, R. Rosengren, L. Safe, and S. Safe. 1989. 2,2',4,4',5,5'-Hexachlorobiphenyl as a 2,3,7,8-tetrachlorodibenzo-p-dioxin antagonist in C57BL/6J mice. Toxicology and Applied Pharmacology 97:561-571.

Haake, J. M., S. Safe, K. Mayura, and T. D. Philips. 1987. Aroclor 1254 as an antagonist of the teratogenicity of 2,3,7,8-tetrachlorodibenzo-p-dioxin. Toxicology Letters 38:299-306.

Lubet, R. A., F. P. Guengerich, and R. W. Nims. 1990. Induction of alkoxyresorufin metabolism: A potential indicator of environmental contamination. Archives of Environmental Contamination and Toxicology 19:157-163.

Neubert, R., L. Maskow, J. Webb, U. Jacob-Muller, A. C. Nogueira, I. Delgado, H. Helge, and D. Neubert. 1993. Chlorinated dibenzo-p-dioxin and dibenzofuran and the human immune system: 1. Blood cell receptor in volunteers with moderately increased body burdens. Life Sciences 53:1995-2006.

Rao, V. R., Y-T. Woo, D. Lai, and J. C. Arcos. 1989. Database on the promoters of chemical carcinogenesis. Environmental Carcinogenesis Reviews C7(2):145-386.

Rao, V. R. 1991a. Molecular mechanism of action of dioxin — Implications on its carcinogenic potency estimation. Toxic Substances Journal 8:335-351.

Rao, V. R. 1991b. Binary effects of carcinogens and tumor promoters — A preliminary chemical structural analysis of BCIDB and PCIDB. Journal of Toxicology and Environmental Health 33:237-248.

Rao, V. R. 1992. Binary combination effects of some pharmacologically active chemicals as promoters of tumorigenesis. Journal of Pharmaceutical Sciences 81:403-407.

Rao, V. R., and A. Unger. 1994a. A novel application of competitive binding model in dioxin risk assessment. Regulatory Toxicology and Pharmacology (communicated).

Rao, V. R., and A. Unger. 1994b. Competitive binding of dioxin congeners with differential affinities for the Ah receptor. Environmental and Toxicological Chemistry (communicated).

Rao, V. R., S. Rheingerover, and A. Unger. 1994c. Application of RISK-PRO and competitive binding model to assess inhalation exposures to dioxins from incinerators. Risk Analysis (under preparation).

Rappe, C., R. Anderson, P-A. Bergqvist, C. Brohede, M. Hansson, L-O. Kjeller, G. Lindstrom, S. Marklund, M. Nygren, S. E. Swanson, M. Tysklind, and K. Wiberg. 1987. Overview on environmental fate of chlorinated dioxins and dibenzofurans. Sources, levels and isomeric pattern in various matrices. Chemosphere 16:1603-1618.

Safe, S. 1990. Polychlorinated biphenyls (PCBs), dibenzo-p-dioxins (PCDDs), dibenzofurans (PCDFs), and related compounds: Environmental and mechanistic considerations which support the development of toxic equivalent f factors (TEFs). Toxicology 21:51-88.

Schecter, A., J. J. Ryan, and J. D. Constable. 1986. Chlorinated dibenzo-p-dioxin and dibenzofuran levels in human adipose tissue and milk samples from North and South Vietnam. Chemosphere 15:1613-1620.

Schecter, A., J. J. Ryan, and J. D. Constable. 1989. Chlorinated dioxins and dibenzofuran in human milk from Japan, India, and the United States of America. Chemosphere 18:975-980.

Stanley, J., K. E. Boggess, J. Onstot, T. Sack, J. C. Remmers, J. Breen, F. W. Kutz, J. Capra, P. Robinson, and G. A. Mack. 1986. PCDDs and PCDFs in human adipose tissue from the EPA FY82 NHATS repository. Chemosphere 15:1605-1612.

U.S. Environmental Protection Agency. 1992a. Biotrol soil washing system for treatment of a wood preserving site — Applications analysis report. EPA/540/A5-91/003, RD 681, EPA/ORD, Cincinnati, OH.

U.S. Environmental Protection Agency. 1992b. National study of chemical residues in fish (Volumes I and II). EPA/823-R-92-008a. EPA/OST, Washington, D.C.

Weber, L.W., M. Lebofsky, B. Stahl, A. Kettrup, and K. Rozman. 1992. Comparative toxicity of four chlorinated dibenzo-p-dioxins (CDDs) and their mixture. Part III: Structure-activity relationships with increased plasma tryptophan levels, but no relationship to hepatic ethoxyresorufin -o-deethylase activity. Archives of Toxicology 66:484-488.

Biomarkers and Occupational Health: Progress and Perspectives. 1995. Pp. 238-256

Xenobiotic-Metabolizing Enzymes in Biomarker Research

Frank J. Gonzalez

A large number of enzymes exist apparently for the sole purpose of metabolizing foreign chemicals (or xenobiotics). It is believed that these enzymes evolved primarily to inactivate or eliminate chemicals found in dietary sources. Plants, for example, produce certain chemicals that are toxic, and animals have enzymes that can degrade and inactivate these toxins. Xenobiotic-metabolizing enzymes have historically been grouped into two categories, the phase I or functionalizing enzymes and the phase II conjugating enzymes (shown in Tables 1 and 2, respectively). The cytochrome P450 and flavin-containing monooxygenases are the major phase I enzymes, while N-acetyltransferases, sulfotransferases, glutathione S-transferases, UDP-glucuronosyltransferases and epoxide hydratases are among the primary phase II enzymes. The xenobiotic-metabolizing enzymes function to inactivate, and in some cases, activate therapeutically used drugs and, in this capacity, they are of tremendous importance to the pharmaceutical industry. Interindividual differences in their expression and drug interactions due to overlapping metabolism of two or more drugs by the same enzyme form can severely compromise drug therapy. Marked species differences in the xenobiotic-metabolizing enzymes also complicate drug safety evaluations. Another important property of these enzymes is their ability to be induced by xenobiotics, many of which are also substrates. It is the interindividual differences in levels of expression of the xenobiotic-metabolizing enzymes and their abilities to be induced by environmental contaminants and dietary chemicals that render them important in the field of biomarker research and development.

P450 cytochromes are the major enzymes involved in drug, and, in particular, carcinogen metabolism (Gonzalez, 1988; 1992; 1994b). P450s exist as a large superfamily of proteins that are classified based on their primary amino acid sequence similarities (Nelson et al., 1993). In mammals, several P450s are involved in specific reactions of steroid biosynthesis and their expression is critical for survival. The vast majority of P450s found in the CYP1, CYP2, CYP3 and CYP4 families metabolize xenobiotics (Gonzalez and Gelboin, 1993). These enzymes exhibit a high degree of species differences, and for this reason, human P450s have been directly studied in recent years (Gonzalez, 1992). A partial list of human P450s that have been found to have high activities toward known classes of carcinogens and mutagens is shown in Table 3.

TABLE 1. PHASE I ENZYMES

Enzymes	Cosubstrates	Multiple Forms
Alcohol dehydrogenases	$O_2/2H$	yes
Aldehyde dehydrogenases	$O_2/2H$	yes
Aldehyde oxidases	$O_2/2H$	yes
Cytochromes P450	$O_2/2H$	yes
Flavin-dependent		
monooxygenases	$O_2/2H$	yes
Monoamine oxidase	$O_2/2H$	no
Myeloperoxidase	H_2O_2	no
Nitric oxide synthases	2H	yes
S-Oxidase		no
Xanthine oxidase	O_2	no
Amidases	H_2O	yes
Arylesterases	H_2O	yes
Carboxylesterases	H_2O	yes
Cholinesterases	H_2O	yes
Epoxide hydratases	H_2O	yes
Azoreductases	2H	yes
Nitroreductases	2H	yes
N-Oxide reductases	2H	yes

In addition to their role in drug metabolism, xenobiotic-metabolizing enzymes are also responsible for activating inert chemicals to their electrophilic derivatives capable of binding to cellular macromolecules and causing cell toxicity, death, and transformation. The principal activating enzymes are the P450s, while the inactivating enzymes are the transferases, although with specific procarcinogens/promutagens, the transferases are also involved in activation. Human variability in levels of these enzymes is thought to be responsible for differential susceptibility to chemical carcinogen-associated cancers. Indeed, recent studies have indicated that levels of certain enzymes confer an altered risk for cancer development and progression. Thus, different interindividual levels of certain enzymes may be considered "biomarkers" for cancer risk. Interindividual differences in enzyme levels are frequently due to the existence of genetic polymorphisms. A number of polymorphisms have been found and these can be diagnosed by polymerase chain reaction (PCR) (Gonzalez and Idle, 1994).

TABLE 2. PHASE II ENZYMES

Enzymes	Cosubstrates	Multiple Forms
UDP-Glucuronosyltransferases	Glucuronic acid	yes
Sulfotransferases	Sulfate	yes
N-Acetyltransferases	Acetate	yes
Glutathione S-transferases	Glutathione	yes
N-Acyltransferases	Amino acids	yes
N,O, and S-Methyltransferase	Methyl	yes

Differences in levels of expression of xenobiotic-metabolizing enzymes can also be due to exposures to inducers. This property is another area that is being exploited in biomarker research and development. Environmental contaminants can result in an increase in levels of P450s, notably CYP1A1. In humans, this P450 is markedly elevated in the lungs, lymphocytes, and placentas of smokers. Efforts are under way to determine whether CYP1A1 expression levels are also correlated with lower level exposure to environmental and industrial chemicals such as the polycyclic aromatic hydrocarbons, polychlorinated biphenyls (PCBs), and related compounds.

Environmental contamination can be monitored by analysis of CYP1A1 expression in fish and rodents. Studies have shown that fish in polluted waters have increased CYP1A1 and, in some cases, liver tumors. Wild mice found at sites contaminated with PCBs were shown to have high levels of these chemicals and increased CYP1A1 activities.

In conclusion, environmental contamination of certain potentially harmful chemicals can be monitored by the induction of metabolic enzymes. Interindividual variation in levels of xenobiotic activating/inactivating enzymes may determine risk or susceptibility for chemical associated diseases in humans.

TABLE 3. MAJOR HUMAN CARCINOGEN/MUTAGEN-METABOLIZING P450s

CYP1A1	Polycyclic aromatic hydrocarbons
CYP1A2	Food mutagens, aflatoxins
CYP2A6	Low molecular weight nitrosamines
CYP2E1	Numerous low molecular weight cancer suspects (acrylonitrile, benzene, nitrosamines, vinyl halides)
CYP3A4	Aflatoxins, food mutagens, nitroaromatic hydrocarbons

BIOMARKERS FOR CANCER RISK

Genetic Polymorphisms

Since xenobiotic enzymes are responsible for either the activation or inactivation of chemical carcinogens, it has not been overlooked that their cellular levels may be associated with risk for cancer development (Idle et al., 1992). Genetic polymorphisms exist in a number of phase I and phase II enzymes (Figure 1). Genetic polymorphisms can be determined by either using a metabolic probe drug or by a genotyping assay such as PCR (Gonzalez and Idle, 1994). Individuals lacking or having low levels of a carcinogen-activating P450 or a carcinogen-inactivating phase II conjugating enzyme would be expected to be at increased risk for cancer development.

CYP2D6 and Lung Cancer

The association of cancer risk with levels of xenobiotic-metabolizing enzymes has been investigated. The CYP2D6 genetic polymorphism has been the most extensively investigated with mixed results (Gonzalez, 1994c). This polymorphism can be determined by genotyping assays, and it affects about 7.5 to 10% of Caucasians who possess two mutant CYP2D6 alleles. In some studies, individuals lacking expression of the enzyme due to the presence of mutant or variant alleles were found to be at low risk for smoking-associated lung cancer. Other case-control studies found no difference between the percentage of deficient subjects in the cancer and control populations. Virtually all studies published to date contain some flaw in their experimental design and therefore the results and conclusions must be viewed cautiously (London et al., 1994). In order to determine whether the CYP2D6 polymorphism is associated with cancer risk, larger and better controlled studies should be designed in which genotyping tests are used to diagnose the presence of mutant alleles.

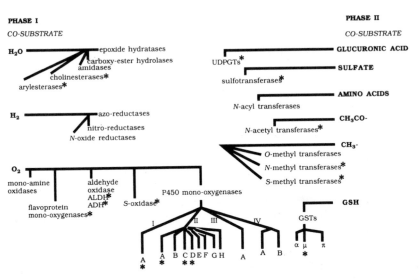

FIGURE 1. Phase I and phase II xenobiotic-metabolizing enzymes. Those in which polymorphisms are known or suspected are denoted by an asterix. Figure reproduced from Idle et al. (1992) with permission from the authors.

CYP1A1 and Lung Cancer

The association of an allele of CYP1A1 and lung cancer in smokers has also been investigated. CYP1A1 is the principal enzyme for metabolic activation of polycyclic aromatic hydrocarbons (Nebert, 1989) and is highly induced in lung tissues of smokers (McLemore et al., 1990).

It was shown in early studies that the extent of CYP1A1 inducibility was greater in lymphocytes derived from lung cancer patients (Kouri et al., 1982). Although a direct relationship between induction in lymphocytes and lung is assumed from animal studies, it has not been demonstrated in humans. A correlation was found between cigarette smoking-associated lung cancer and an CYP1A1 allele encoding a P450 with an Ile → Val amino acid change in Japanese (Kawajiri et al., 1993). The molecular basis of this association is unknown since the catalytic activity or inducibility of this minor allele has not been examined. The frequency of this allele is lower in Caucasians and in a limited study of Norwegian cancer patients, there was no difference in allele frequency between lung cancer patients and matched controls (Tefre et al., 1991).

GSTM1 and Lung Cancer

A genetic polymorphism exists for a glutathione S-transferase GST-M1 in which about half the population is deficient in this enzyme due to a deletion of the gene (Seidergard et al., 1985; 1988). Since this enzyme is able to deactivate carcinogens such as arene oxides and polycyclic aromatic hydrocarbons, a deficiency could result in cancer susceptibility. As with the other cancer associations, the role of GSTM1 in cancer remains controversial. An early study of 66 lung cancer patients and controls revealed an underrepresentation of the active gene in patients (Seidergard et al., 1986). The association was confirmed in a study of 176 Japanese patients (Kihara et al., 1994). In nonadenocarcinoma patients, the frequency of the null genotype was about 64% compared to 48% in matched controls. In adenocarcinoma patients, the frequency of the deficiency was 54%, suggesting that only smoking-associated cancers were affected by these polymorphisms. This was confirmed by a study of the extent of smoking; the proportion of GSTM1 null genotype was found to increase to 75% in patients with the highest smoking index (Table 4). A similar association of the null alleles with lung cancer in heavy smokers was found by others (Nazar-Stewart et al., 1993).

In contrast to these results, others have not found a difference in the percentage of deficient subjects between lung cancer patients and controls (Heckbert et al., 1992; Brockmoller et al., 1993). However, an increased risk was also found associated with the null genotype of GSTM1 for colon cancer patients (Zhong et al., 1993).

By analysis of P450 and GSTM1 genotypes, a strong association with lung cancer risk was found in patients homozygous for the rare

CYP1A1 allele and the null GSTM1 suggesting that both P450s and transferases are critical for human carcinogenesis (Hayashi et al., 1992). Follow-up confirmation studies by other groups have not been conducted.

TABLE 4. PERCENTAGE OF HOMOZYGOUS NULL GSTM1 GENOTYPES IN MALE SUBJECTS AS A FUNCTION OF TOBACCO SMOKE EXPOSURE[1]

| | Smoking Index[2] | | | |
	<800	800-1200	≥1200	Total
Kreyberg I	46	60	73	60
Squamous cell	50	55	72	60
Small cell	51	67	75	60
Kreyberg II				
Adenocarcinoma	52	55	50	53
Control	45	43	48	45

[1] Data taken from Kihara et al., 1994
[2] Sum of cigarettes smoked per day × years of smoking

N-Acetyltransferase 2 and Cancer Risk

The N-acetyltransferases (NAT) have been shown to be involved in both inactivation and activation pathways of carcinogen metabolism (Figure 2). The acetylation of amino groups results in an inactive metabolite, while O-acetylation of an N-hydroxy group of certain arylamine and heterocyclic arylamine carcinogens leads to an active electrophilic derivative. Thus, it might be expected that the association of the NAT enzyme with cancer susceptibility might be dependent on the type of carcinogen.

N-Acetyltransferase 2 (NAT2) is polymorphic in humans and about 50% of the population is deficient in the enzyme as a result of mutant genes (Grant et al., 1992). Early evidence showed an association between the deficiency and occupational bladder cancer (Cartwright et al., 1982). Others, using both phenotyping and genotyping to determine the NAT2 alleles, found no association between the deficiency and increased risk of bladder cancer in workers occupationally exposed to benzidine, a carcinogen that can be inactivated by NAT2 (Hayes et al., 1993).

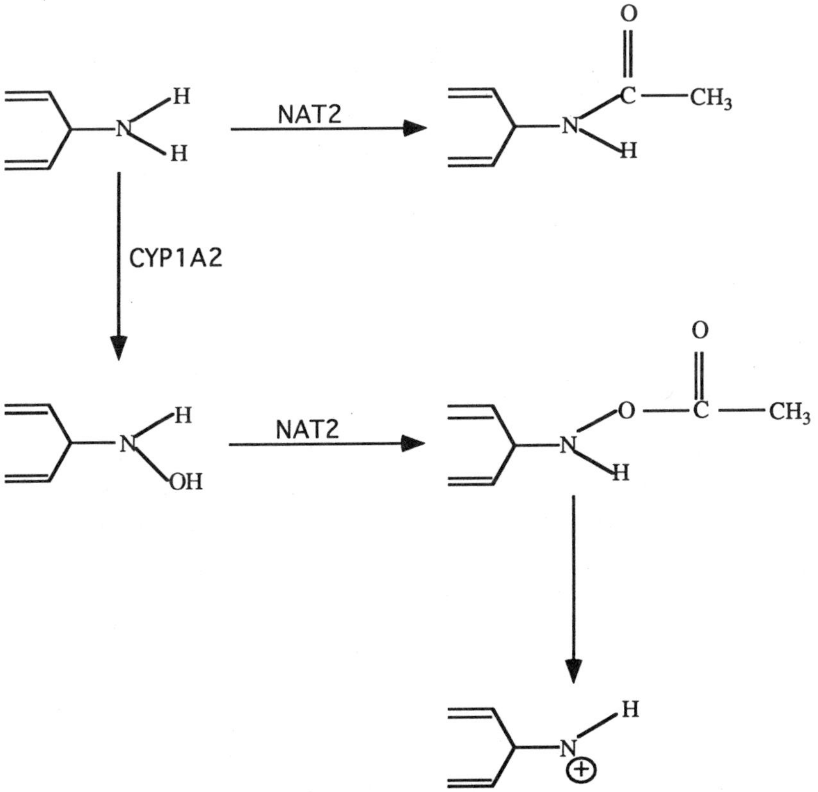

FIGURE 2. Role of NAT2 in metabolism of arylamines. The enzyme can carry out both N-acetylation inactivation pathways and O-acetylation activation pathways. The N-acetoxy metabolite may also be hydroxylated by P450 which, followed by migration of the acetyl group to the oxygen, can also lead to the active carbonium ion.

In a study of smokers, the levels of 3- and 4-aminobiphenyl (ABP) hemoglobin adducts, a biomarker for recent activation of ABP, were higher in slow acetylators having deficient NAT2 alleles, showing a direct correlation between the polymorphism and a biomarker for carcinogen activation (Yu et al., 1994). Higher levels of ABP adducts were found in smokers. The finding was confirmed in Caucasians, Asians, and Blacks. The latter race had the lowest proportion of NAT2-deficient subjects and the highest level of adducts. Blacks also have the highest frequency of bladder cancer, suggesting a causal association of NAT2 genotype with cancer risk. This study was extended to assess the relationship between NAT2 genotype and DNA adducts (Vineis et al., 1994). Slow acetylators were found to contain higher levels of aminobiphenyl-DNA adducts than rapid acetylators. The effect of NAT2 genotype on both ABP and the AB-DNA adducts became less significant in smokers. These studies suggest that NAT2 genotype plays a role in carcinogen activation as measured by protein and DNA adducts under conditions of low dose environmental exposure.

The role of NAT2 in colon cancer has also been addressed. An increased risk was found in individuals having rapid acetylation activity and high CYP1A2 activity (Minchin et al., 1993). This association is opposite that found in bladder cancer indicating a role of NAT2 in carcinogen activation. CYP1A2 is the primary enzyme responsible for N-hydroxylation of arylamine carcinogens and heterocyclic arylamine food mutagens. The N-hydroxy metabolite must be esterified by acetate or sulfate in order to form the proximate metabolite capable of binding to DNA. These data would suggest that the N-hydroxy metabolite is formed in the liver by CYP1A2 and is transported to the colon where it is activated by NAT2. Others found no difference in genotype frequency and colon cancer association in Whites or Blacks and even suggested that NAT1, which is not subject to genetic polymorphism, is the only form expressed in human colon (Rodriguez et al., 1993). Further studies are warranted to determine the association of NAT2 genotype with colon cancer.

In nonsmokers, 4-aminobiphenyl DNA adducts can be used as a biomarker for both exposure to 4-aminobiphenyl and for NAT2 genotype. Aflatoxin B1 (AFB) DNA adducts in urine are a biomarker for exposure to this human hepatocarcinogen (Groopman et al., 1994). These adducts were found to be higher in patients that ultimately get liver cancer. AFB is metabolically activated by cytochromes P450. Although several human P450s can produce the active 8,9-epoxide metabolite (Aoyama et al., 1990 a,b), CYP1A2 appears to be the P450 with the lowest Michaelis-Menten constant K_m for carrying out this reaction (Crespi et al., 1991; Gallagher et al., 1994). A form of glutathione S-transferase is capable of conjugating and inactivating the epoxide and studies in rats have shown that induction of GST by the drug oltipraz protects rats from AFB liver cancer (Kensler et al., 1987). Perhaps this drug, which also decreases AFB-DNA adducts in AFB fed rats, would be of use in chemoprevention of hepatocarcinogenesis in areas of China and Africa where this mold toxin is a serious contaminant of grains.

Estradiol Hydroxylase Activity and Breast Cancer Risk

Elevated estradiol-17B 16α-hydroxylase activity in terminal duct lobular units of human breast was found to be associated with increased risk of breast cancer (Osborne et al., 1993). These are the presumed target sites of breast carcinogenesis. A four-fold to five-fold difference was found in activity between controls and cancer patients. Despite the age difference between the groups and the small sample size, these data could indicate a very important enzymatic biomarker for breast cancer risk. A polymorphism for this activity in humans has not been demonstrated, nor has the P450 form involved in this oxidation been identified. The possible mechanisms by which increased 16α-hydroxylase could be associated with increased cancer risk have been discussed (Nebert, 1993). A P450 that metabolizes carcinogens could be involved since it is well established that a P450 form capable of metabolic activation of chemical carcinogens can also metabolize steroids and other chemicals to stable metabolites. For example, CYP1A2, which metabolizes numerous carcinogens and mutagens (Table 3) is an estrogen 2-hydroxylase (Aoyama et al., 1990 a,b).

BIOMARKERS FOR CARCINOGEN EXPOSURE

Human Exposure

As noted earlier, 4-aminobiphenyl hemoglobin adducts and aflatoxin B1 DNA adducts have been proposed as biomarkers for carcinogen exposure. These biomarkers reflect recent exposure and their predictive value for cancer cannot be assessed easily since cancer initiation in humans is likely to be followed by a long dormant period of perhaps 20 years. These biomarkers can, however, be used to determine current environmental exposures.

Induction of CYP1A1 and CYP1A2 activity has been proposed as a method to determine exposure to polycyclic aromatic hydrocarbon and dioxin exposures. The mechanism for induction by this class of chemicals has been well studied (Gonzalez, 1994a). The heterodimeric Ah receptor is required for binding the inducer ligand and transmitting the signal for transcriptional activation to the nucleus. The receptor complex binds to a regulatory element upstream of target genes and, through a mechanism that is not completely understood, activates transcription.

The caffeine breath test was first developed to measure CYP1A2 activity in humans (Lambert et al., 1986). In this test, the N-3 methyl group of caffeine is labeled and after it is demethylated by the hepatic P450, it can be measured as expired carbon dioxide. By using this assay, higher levels of activity were detected in a cohort of Michigan fishermen who were exposed to polybrominated biphenyls as compared with matched controls (Lambert et al., 1990).

In another study, CYP1A1 gene expression was monitored in human lymphocytes of railroad workers exposed to creosote in order to determine exposure to polycyclic aromatic hydrocarbons (Cosma et al., 1992). This was accomplished by measuring mRNA levels. As compared with matched controls, workers had a two-fold increase in CYP1A1 mRNA only when lymphocytes were collected in the summer (Table 5). No significant differences were found between matched controls and workers when activity was measured in the fall and winter. The level of increase, although quite small as compared with fully-induced lymphocytes (Jaiswal et al., 1985), suggests that the polycyclic aromatic hydrocarbons in the creosote volatilize in summer and contribute to worker exposure. It is unclear how well the CYP1A1 mRNA induction assays reflect the extent of exposure.

A reverse transcriptase-polymerase chain reaction method was developed to measure low levels of CYP1A1 mRNA (Vanden Heuvel et al., 1993). By this method mRNA could be readily measured in unstimulated lymphocytes. The high level of sensitivity may be of use in measuring CYP1A1 expression in exposed populations.

Rodents as Biomarkers for Soil Contamination

Induction of CYP1A1 activity has been used to determine exposure of mice at polychlorinated biphenyl (PCB) contaminated reference sites (Lubet et al., 1992). The levels of activities in individual mice were correlated with hepatic PCB burdens. The increased activities were also reflected in higher levels of CYP1A1 protein. These data suggest that indigenous induction of mouse CYP1A1 may be employed as sensitive biomarkers of environmental exposure to PCBs. Based on laboratory studies it was suggested that feral rats may be even more sensitive than mice for monitoring PCB exposure (Novak and Qualls, 1989; Lubet et al., 1992).

TABLE 5. EXPRESSION OF THE CYP1A1 GENE IN PERIPHERAL LYMPHOCYTES
OF CREOSOTE EXPOSED RAILROAD WORKERS[1]

Season	mRNA Level		
	Controls	Worker	Workers/Controls
Summer	3.1	5.7	1.9
Fall	3.1	3.8	1.2
Winter	4.9	3.7	0.8

[1] Reproduced from Cosma et al., 1992

Fish as Biomarkers for Water Contamination

CYP1A1 is inducible by polycyclic aromatic hydrocarbons (PAH) and PCBs (Stegeman, 1989), and fish have been used as biological monitors for PAH and PCB contamination (Payne et al., 1987; Goksoyr and Forlin, 1992). A number of studies have been conducted using fish to monitor water pollution levels of CYP1A1 mRNA in livers of Atlantic tomcod collected from two sites in the Hudson River in which the pollution levels were higher than those found in a river in Maine (Kreamer et al., 1991). Placing the Hudson River fish in an aquarium resulted in a loss of mRNA content which reached basal levels in 5 days. The fish could be re-induced by laboratory exposure to sediment from a contaminated site. These studies demonstrate the ability of this species to serve as an environmental monitor of aquatic pollution.

Studies have also been carried out analyzing Tilapia, a hearty fish that is able to live in the heavily polluted Damsui River in northern Taiwan. Fish collected in a particularly polluted upstream region of the river had higher levels of CYP1A1 (Figure 3) and its associated activities than Tilapia collected at a non-polluted downstream site (Ueng et al., 1992; 1994). Administration to fish in a laboratory aquarium of sediment taken from polluted sites of the Damsui also markedly induce CYP1A1 activity.

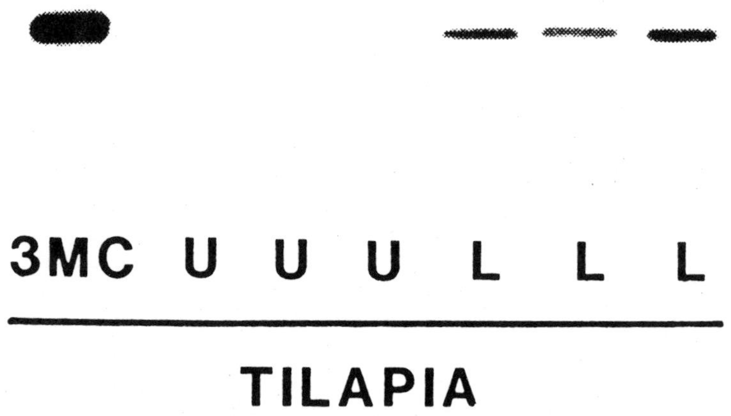

3MC U U U L L L

TILAPIA

FIGURE 3. Western blot analysis of the CYP1A1 protein in livers of the fish Tilapia treated with 3-methylcholanthrene (3MC) or harvested from upstream clean water (U) or downstream dirty water (L). Data taken from Ueng et al. (1992) with permission from the authors.

Fish have been used to monitor inducers in paper mill effluents (Mather-Mihaich and DiGiulio, 1991; Adams et al., 1992). CYP1A1 was found to be induced by as little as a 10% diluted effluent (Mather-Mihaich and DiGiulio, 1991). Inducers found in effluent collected in April were more potent than those collected in August. The chemicals in the effluent causing the inducing effects are unknown.

REFERENCES

Adams, S. M., W. D. Crumby, M. S. Greeley, L. R. Shugart, and C. F. Saylor. 1992. Responses of fish populations and communities to pulp mill effluents: a holistic assessment. Ecotoxicology and Environmental Safety 24:347-360.

Aoyama, T., K. Korzekwa, J. Gillette, H. V. Gelboin, and F. J. Gonzalez. 1990a. Estradiol metabolism by complementary deoxyribonucleic acid-expressed human cytochrome P450s. Endocrinology 126:3101-3106.

Aoyama, T., S. Yamano, P. S. Guzelian, H. V. Gelboin, and F. J. Gonzalez. 1990b. Five of 12 forms of vaccinia virus-expressed human hepatic cytochrome P450 metabolically activate aflatoxin B1. Proceedings of the National Academy of Sciences (USA) 87:4790-4793.

Brockmoller, J., R. Kerb, N. Drakoulis, M. Nitz, and I. Roots. 1993. Genotype and phenotype of glutathione S-transferase class mu isoenzymes mu and psi in lung cancer patients and controls. Cancer Research 53:1004-1011.

Cartwright, R. A., R. Glashan, H. J. Rogers, R. A. Ahmad, D. Barham-Hall, E. Higgins, and M. Kahn. 1982. Role of N-acetyltransferase phenotypes in bladder carcinogenesis: a pharmacogenetic epidemiological approach to bladder cancer. Lancet ii:842-845.

Cosma, G. N., P. Toniolo, D. Currie, B. S. Pasternack, and S. J. Garte. 1992. Expression of the CYP1A1 gene in peripheral lymphocytes as a marker of exposure to creosote in railroad workers. Cancer Epidemiology and Biomarker Prevention 1:137-142.

Crespi, C. L., B. W. Penman, D. T. Steimel, H. V. Gelboin, and F. J. Gonzalez. 1991. The development of a human cell line stably expressing human CYP3A4: role in the metabolic activation of aflatoxin B1 and comparison to CYP1A2 and CYP2A3. Carcinogenesis 12:355-359.

Gallagher, E. P., L. C. Wienkers, P. L. Stapleton, K. L. Kunze, and D. L. Eaton. 1994. Role of human microsomal and human complementary DNA-expressed cytochromes P4501A2 and P4503A4 in the bioactivation of aflatoxin B1. Cancer Research 54:101-108.

Goksoyr, A., and L. Forlin. 1992. The Cytochrome P-450 system in fish aquatic toxicology and environmental monitoring. Aquatic Toxicology 22:287-311.

Gonzalez, F. J. 1988. The molecular biology of cytochrome P450s [published erratum appears in Pharmacological Reviews 1989 March 41:91-2]. Pharmacological Reviews 40:243-288.

Gonzalez, F. J. 1992. Human cytochromes P450: problems & prospects. Trends in Pharmacological Sciences 13:346-352.

Gonzalez, F. J. 1994a. In: Skin Cancer: Mechanisms and Human Relevance, H. Mukhtar, Ed., CRC Press, Boca Raton, FL. (In press).

Gonzalez, F. J. 1994b. In: Principle of Cancer Drug Pharmacology, G.A. Milano and M.J. Ratain, Eds. Marcel Dekker Inc., New York. (In press).

Gonzalez, F. J. 1994c. In: Cytochromes P450: Metabolic and Toxicological Aspect, C. Ioannides, Ed., CRC Press, Boca Raton, FL. (In press).

Gonzalez, F. J., and H. Gelboin. 1993. Role of human cytochrome P-450s in risk assessment and susceptibility to environmentally based disease. Journal of Toxicology and Environmental Health 40:289-308.

Gonzalez, F. J., and J. R. Idle. 1994. Pharmacogenetic phenotyping and genotyping. Present status and future potential. Clinical Pharmacokinetics 26:59-70.

Grant, D. M., M. Blum, and U. A. Meyer. 1992. Polymorphisms of N-acetyltransferase genes. Xenobiotica 22:1073-1081.

Groopman, J. D., G. N. Wogan, B. D. Roebuck, and T. W. Kensler. 1994. Molecular biomarkers for aflotoxins and their application to human cancer prevention. Cancer Research 54:1907s-1911s.

Hayashi, S., K. Watanabe, and K. Kawajiri. 1992. High susceptibility to lung cancer analyzed in terms of combined genotypes of P450IA1 and Mu-class glutathione S-transferase genes. Japanese Journal of Cancer Research 83:866-870.

Hayes, R. B., N. Rothman, F. Broly, N. Caporaso, P. Feng, X. You, S. Yin, R. L. Woosley, and U.A. Meyer. 1993. N-Acetylation phenotype and genotype and risk of bladder cancer in benzidine-exposed workers. Carcinogenesis 14:675-678.

Heckbert, S. R., N. S. Weiss, S. K. Hornung, D. L. Eaton, and A. Motulsky. 1992. Glutathione S-transferase and epoxide hydrolase activity in human leukocytes in relation to risk of lung cancer and other smoking-related cancers. Journal of the National Cancer Institute 84:414-422.

Idle, J. R., M. Armstrong, A. V. Boddy, C. Boustead, S. Cholerton, J. Cooper, A. K. Daly, J. Ellis, W. Gregory, H. Hadidi, C. Hofer, J. Holt, J. Leathart, N. McCracken, S. O. Monkman, J. E. Painter, H. Taber, D. Walker, and M. Yule. 1992. The pharmacogenetics of chemical carcinogenesis. Pharmacogenetics 2:246-258.

Jaiswal, A. K., F. J. Gonzalez, and D. W. Nebert. 1985. Human P1-450 gene sequence and correlation of mRNA with genetic differences in benzo[a]pyrene metabolism. Nucleic Acids Research 13:4503-4520.

Kawajiri, K., K. Nakachi, K. Imai, J. Watanabe, and S. Hayashi. 1993. The CYP1A1 gene and cancer susceptibility. Critical Reviews of Oncology and Hematology 14:77-87.

Kensler, T. W., P. A. Enger, P. H. Donan, J. D. Groopman, and B. D. Roebuck. 1987. Mechanism of protection against aflatoxin tumorigenicity in rats fed 5-(2-pyrazinyl)-4-methyl-1,2-dithiol-3-thione (oltipraz) and related 1,2-dithiol-3-thiones and 1,2-dithiol-3-ones. Cancer Research 47:4271-4277.

Kihara, M., M. Kihara, and K. Noda. 1994. Lung cancer risk of GSTM1 null genotype is dependent on the extent of tobacco smoke exposure. Carcinogenesis 15:415-418.

Kouri, R. E., C. E. McKinney, D. J. Slomiany, D. R. Snodgrass, N. P. Wray, and T. L. McLemore. 1982. Positive correlation between high aryl hydrocarbon hydroxylase activity and primary lung cancer as analyzed in cryopreserved lymphocytes. Cancer Research 42:5030-5037.

Kreamer, C. L., K. Squibb, D. Gioeli, S. J. Garte, and I. Wirgin. 1991. Cytochrome P450IA mRNA expression in feral Hudson River tomcod. Environmental Research 55:64-78.

Lambert, G. H., D. A. Schoeller, A. N. Kotake, C. Flores, and C. Hay. 1986. The effect of age, gender, and sexual maturation on the caffeine breath test. Developments in Pharmacological Theraputics 9:375-388.

Lambert, G. H., D. A. Schoeller, H. E. Humphrey, and A. N. Kotake. 1990. The caffeine breath test and caffeine urinary metabolite ratios in the Michigan cohort exposed to polybrominated biphenyls: a preliminary study. Environmental Health Perspectives 89:175-181.

London, S. J., A. K. Daly, D. C. Thomas, N. E. Caporaso, and J. R. Idle. 1994. Methodological issues in the interpretation of studies of the CYP2D6 genotype in relation to lung-cancer risk. Pharmacogenetics 4:107-108.

Lubet, R. A., R. W. Nims, L. E. Beebe, S. D. Fox, H. J. Issaq, and K. McBee. 1992. Induction of hepatic CYP1A activity as a biomarker for environmental exposure to Aroclor 1254 in feral rodents. Archives of Environmental Contamination and Toxicology 22:339-344.

Mather-Mihaich, E., and R. T. DiGiulio. 1991. Oxidant, mixed-function oxidase and peroxisomal responses in catfish exposed to a bleached kraft mill effluent. Archives of Environmental Contamination and Toxicology 20:391-397.

McLemore, T., S. Adelberg, M. C. Lim, N. A. McMahon, S. J. Yu, W. C. Hubbard, M. Czerwinski, B. P. Coudert, J. A. Moscow, S. Stinson, R. Storeny, R. A. Lubet, J. C. Eggleston, M. R. Boyd, and R. W. Hines. 1990. Expression of CYP1A1 gene in patients with lung cancer: evidence for cigarette smoke-induced gene expression in normal lung tissue and for altered gene regulation in primary pulmonary carcinomas. Journal of the National Cancer Institute 82:1333-1339.

Minchin, R. F., F. F. Kadlubar, and K. F. Ilett. 1993. Role of acetylation in colorectal cancer. Mutation Research 290:35-42.

Nazar-Stewart, V., A. G. Motulsky, D. L. Eaton, E. White, S. K. Hornung, Z. T. Leng, P. Stapleton, and N. S. Weiss. 1993. The glutathione S-transferase mu polymorphism as a marker for susceptibility to lung carcinoma. Cancer Research 53:2313-2318.

Nebert, D. W. 1989. The Ah locus: genetic differences in toxicity, cancer, mutation, and birth defects. Critical Reviews in Toxicology 20:153-174.

Nebert, D. W. 1993. Elevated estrogen 16 alpha-hydroxylase activity: is this a genotoxic or nongenotoxic biomarker in human breast cancer risk? [editorial; comment] Journal of the National Cancer Institute 85:1888-1891.

Nelson, D. R., T. Kamataki, D. J. Waxman, F. P. Guengerich, R. W. Estabrook, R. Feyereisen, F. J. Gonzalez, M. J. Coon, I. C. Gunsalus, O. Gotoh, K. Okuda, and D. W. Nebert. 1993. The P450 superfamily: update on new sequences, gene mapping, accession numbers, early trivial names of enzymes, and nomenclature. DNA Cellular Biology 12:1-51.

Novak, J., and C. W. Qualls. 1989. Effects of phenobarbital and 3-methylcholanthrene on the hepatic cytochrome P-450 metabolism of various alkoxyresorufin ethers in the cotton rat (*Sigmodon hispidus*). Comparative Biochemistry and Physiology 94C:543-545.

Osborne, M. P., H. L. Bradlow, G. Y. C. Wong, and N. T. Telang. 1993. Upregulation of estradiol C16 alpha-hydroxylation in human breast tissue: a potential biomarker of breast cancer risk [see comments]. Journal of the National Cancer Institute 85:1917-1920.

Payne, J. F., L. L. Fancey, A. D. Rahimtula, and E. L. Porter. 1987. Review and perspective on the use of mixed-function oxygenase enzymes in biological monitoring. Comparative Biochemistry and Physiology 86C:233-245.

Rodriguez, J. W., W. G. Kirlin, R. J. Ferguson, and M. A. Doll. 1993. Human acetylator genotype: relationship to colorectal cancer incidence and arylamine N-acetyltransferase expression in colon cytosol. Toxicology 67:445-452.

Seidergard, J., J. W. DePierre, and R. W. Pero. 1985. Hereditary interindividual differences in the glutathione transferase activity towards trans-stilbene oxide in resting human mononuclear leukocytes are due to a particular isozyme(s). Carcinogenesis 6:1211-1216.

Seidergard, J., R. W. Pero, D. B. Miller, and E. J. Beattie. 1986. A glutathione transferase in human leukocytes as a marker for the susceptibility to lung cancer. Carcinogenesis 7:751-753.

Seidergard, J., W. R. Vorachek, R. W. Pero, and W. R. Pearson. 1988. Hereditary differences in the expression of the human glutathione transferase active on trans-stilbene oxide are due to a gene deletion. Proceedings of the National Academy of Sciences (USA) 85:7293-7297.

Stegeman, J. J. 1989. Cytochrome P450 forms in fish: catalytic, immunological and sequence similarities. Xenobiotic 19:1093-1110.

Tefre, T., D. Ryberg, A. Haugen, D. W. Nebert, V. Skaug, A. Brogger, and A. L. Borresen. 1991. Human CYP1A1 (cytochrome P(1)450) gene: lack of association between the Msp I restriction fragment length polymorphism and incidence of lung cancer in a Norwegian population. Pharmacogenetics 1:20-25.

Ueng, T. H., Y. F. Ueng, and S. S. Park. 1992. Comparative induction of cytochrome P-450-dependent monooxygenases in the livers and gills of tilapia and carp. Aquatic Toxicology 23:49-64.

Ueng, Y. F., T. Y. Liu, and T. H. Ueng. 1994. Environmental Contamination and Toxicology (In press).

Vanden Heuvel, J. P., G. C. Clark, C. Thompson, Z. McCoy, C. R. Miller, G. W. Lucier, and D. A. Bell. 1993. CYP1A1 mRNA levels as a human exposure biomarker: use of quantitative polymerase chain reaction to measure CYP1A1 expression in human peripheral blood lymphocytes. Carcinogenesis 14:2003-2006.

Vineis, P., H. Bartsch, N. Caporaso, A. M. Harrington, F. F. Kadlubar, M. T. Landi, C. Malaveille, P. G. Shields, P. Skipper, G. Talaska, and S. Tannenbaum. 1994. Genetically based N-acetyltransferase metabolic polymorphism and low-level environmental exposure to carcinogens. Nature 369:154-156.

Yu, M. C., P. L. Skipper, K. Taghizadeh, S. R. Tannenbaum, K. K. Chan, B. E. Henderson, and R. K. Ross. 1994. Acetylator phenotype, aminobiphenyl-hemoglobin adduct levels, and bladder cancer risk in white, black and Asian men in Los Angeles, California. Journal of the National Cancer Institute 86:712-716.

Zhong, S., A. H. Wyllie, D. Barnes, C. R. Wolf, and N. K. Spurr. 1993. Relationship between the GSTM1 genetic polymorphism and susceptibility to bladder, breast and colon cancer. Carcinogenesis 14:1821-1824.

Biomarkers and Occupational Health: Progress and Perspectives. 1995. Pp. 257-263

Implications of Large Scale DNA Analysis for the Development and Application of Biomarkers

Charles R. Cantor, Takeshi Sano, and Cassandra L. Smith

The Human Genome Project has stimulated the rapid development of new strategies and new tools for DNA analysis. In the first three years of the project, the rate and the efficiency of DNA mapping and sequencing have increased by an order of magnitude. Techniques currently in the early stages of development promise several orders of magnitude further increase in the rate of genome analysis. These new methods are not actually needed to complete the genome project as originally planned. However, they will be necessary if large scale studies of human diversity, routine clinical analysis by DNA sequencing, or routine large scale use of DNA markers to examine environmental insults are ever to become economically feasible.

This paper will discuss the implications of the advances that are already in progress and some of the methods that may be applicable in the near future. The ultimate impact of these advances on the development and use of biomarkers will be monumental. Consider, for example, their potential impact on our ability to survey an entire environment for the presence of microorganisms. The methods being developed for genome research currently provide a means for very rapid analysis of DNA samples from multiple sources. Only minor variations of these methods are needed to turn our attention from intensive exploration of a single genome to wide scale surveys of selected regions of many genomes.

A Computational and Scientific Challenge

An interesting feature of the human population is that everyone is different. Those differences are encoded predominantly in DNA base pairs, and a current estimate is that any two individuals differ at the DNA level at about 6 million locations in their genome.

The earth's population currently is 5 billion, and hopefully, we will someday have the technology to catalog all those differences. That catalog would be a database of 3×10^{16} entries, which is a large database by today's standards. Using today's technology, its storage would require between 1 million and 10 million CD ROMs, which is enormous. Computer storage capability is accelerating at the rate of about 10^4-fold per decade, and if that acceleration continues as projected, in 15 years, the database of all human diversity will be the equivalent of a single CD ROM, which will be trivial from a computer science point of view.

Therefore, what is needed is the technology to attain that size database. Assuming this is possible, it is important to consider the implications and uses of what will be achieved both by the Human Genome Project and by studies of human variability that will follow.

Clearly, new methods will be developed, and they will include amplification methods, sequencing methods, and the ability to deal with an enormous amount of DNA sequence information. Human genes involved in the responses to environmental challenges will be identified, as well as DNA regions that are particularly sensitive to the environment. This information will allow us to use the methods that are now available in the most effective way possible to look for environmentally induced changes.

Currently, as part of the Human Genome Project, all biological databases are being integrated. It is possible to pass smoothly from clinical data to sequence data to genetic data. This will improve even more over the next few years, and it will facilitate any kind of biological inquiry. In addition, there is a large investment in technology development, and there are many unforeseen advantages and consequences of this development.

DNA and DNA Markers

DNA is a unique molecule for analysis, whether it is used as a target or a reagent. There is no other known type of molecule that can be specifically recognized by another molecule with a complementary sequence in such a totally general way. Any DNA sequence can be synthesized and replicated in any amounts desired, thus allowing unrestricted manipulation. Secondly, all individuals in all populations, except for clonal populations, are distinguishable at the DNA level. And finally, as a target, DNA is quite stable. Even fossil material can be studied, although not very well. Therefore, DNA has extremely attractive aspects. It is possible to detect and work with the amplification products of single DNA molecules at a level of sensitivity unmatched by any other system. Other molecules can be tagged with DNA and studied. Finally, we can work with large quantities of DNA simultaneously in various chip or array formats.

DNA markers can be used in at least three important ways. First, it is possible to examine a site and determine the organism, the species, even the individual, and to monitor these aspects as a function of time. DNA can be attached as a tag to track virtually any chemical agent. DNA also serves as an indicator of environmental damage, which has been generally its major use as a biomarker in the past, or for altered expression profiles, which will probably become a more popular use as the sequencing technologies improve.

Large arrays of DNA samples can be prepared either manually or robotically. Thirty thousand-element arrays are currently being routinely prepared by some groups. No other chemical clusters can be studied on this scale with the possible exception of combinatorial antigens. These arrays enable scientists to look at all of their elements simultaneously. DNA is used as a probe to look for complementary DNA. This is a technique called hybridization, and it is the basic technique used in DNA analysis. In principle, multiple-colored tags, including isotopes or fluorescent materials, can be used to conduct many tests simultaneously. DNA arrays can also be used to make arrays of other species. A DNA array can be replicated and it might eventually become relatively inexpensive to manufacture these copies. Arrays can be customized to conduct different analytical tests.

In some cases, it is desirable to immortalize the DNA of the organism under study. If there are just a few cells, or perhaps one cell, that will not grow in the laboratory, it is necessary to amplify them or it in order to study the DNA of that organism. This technique is called whole genome amplification; numerous protocols have been proposed to accomplish this process, but they do not work very well (Kinzler and Vogelstein, 1989; Ludecke et al., 1989; Zhang et al., 1992). There is, however, a technique that works well and is potentially useful in a wide range of settings. The method is called tagged random primer PCR (Grothues et al., 1993), and it is based on the polymerase chain reaction which is a way of replicating DNA *in vitro* (Figure 1).

A chimeric piece of DNA is prepared containing a constant sequence, which we create, and with all 4^9 possible sequence combinations at the end. A DNA target is then amplified with this chimeric primer for a couple of cycles, and the high molecular weight DNA material is purified. The primer is then replaced with a new primer, which now exactly corresponds to the unique sequence. Numerous additional cycles of amplification are then carried out until there is enough material for further work.

Therefore, the first step is amplification of everything, and the second step is a very clean amplification that actually results in a sufficient amount of product. In an amplification experiment that started with material equivalent to only three human chromosomes, which is approximately 10 percent of a human cell, we were still able to amplify enough DNA for use with fluorescent staining.

In another experiment to determine whether we could amplify a whole genome at random, we took 900 clones containing fragments of an organism of approximately 15 million base pairs. We used a DNA probe, which was an electrophoretically purified fraction consisting of a single chromosome from that organism, to interrogate the array and determine which of these clones contained that particular chromosome. To test the amplification system, we then took a small part of the purified chromosome preparation, amplified it using this random amplification system, and then interrogated the array with the amplification product. Almost all portions of this chromosome were uniformly amplified.

DNA arrays can be used as sequence-specific detectors, but, in fact, arrays are really more useful in sequence-specific capture. After capturing a target on an element of an array, additional steps can be used with that captured target. Currently, we want to sequence large numbers of captured targets and read out DNA sequence information, not of just a few molecules, but of many molecules simultaneously.

In order to look at all of the differences in all people, it will require sequencing rates of about 10^7 bases per day before this goal becomes even remotely possible.

Based on a review of past and current sequencing speeds, it is estimated that manual DNA sequencing averages about 100 bases per day. Sophisticated automated equipment can sequence about 20,000-200,000 base pairs per day. Only a few laboratories are currently able to achieve the upper range of that estimate. With improvement of current methods, an ideal rate of about 5 million bases per day is envisioned. The use of hybridization to directly sequence might result in a rate of 2×10^7 base pairs per day. Although these numbers appear to be unrealistically high, it is interesting to note that *E. coli* accurately replicates its own DNA at the rate of 2×10^8 bases per day.

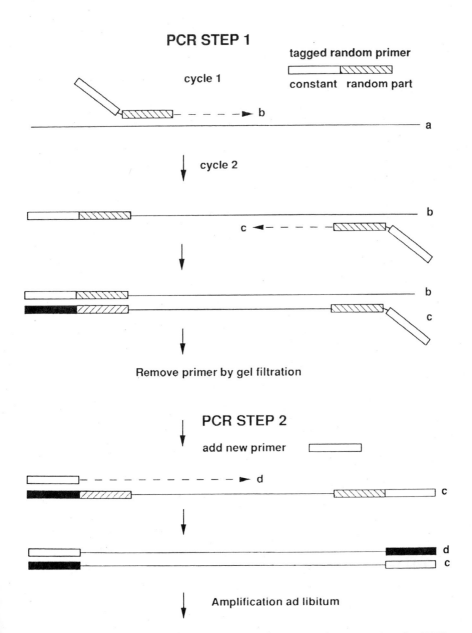

FIGURE 1. The tagged random primer polymerase chain reaction for DNA amplification.

In the ultimate applications, mass spectrometry might result in sequencing rates of about 3×10^9 base pairs per day. Currently, the rate-limiting step is determining the DNA sequence, but in the near future, sample preparation may be the rate-limiting step. Eventually, the identification of samples worth sequencing will be an important factor.

Future Applications

One aspect of our work resulted in an improvement in the technology that can be used to work with antigens or antibodies. It uses the protein streptavidin, which is typically used in clinical or immuno-assays as the coupler between a probe and a specificity molecule. The target might be an antigen, an environmental marker, or possibly an isotope or radioisotope. For improved sensitivity, we made a chimera between streptavidin and staphylococcal protein A via genetic engineering. The resulting chimera binds four antibodies through its staph A portion, because protein A is specific for the constant portion of antibodies. Thus, it allowed us to make a link between DNA and the world of antigens and antibodies.

A specific antibody will detect an epitope, but the ability to detect DNA is much better than the ability to detect antibodies or antigens. Therefore, DNA technology, through amplification with a polymerase chain reaction can be used to detect the presence of an antigen (Sano et al., 1992; Sano, 1993). This is done in the laboratory by immobilizing the antigen on a surface, blocking the surface, binding the antibody, binding this detection system, and then doing the amplification. In the early forms of this technology, electrophoresis was used to look at the products. That is no longer done (Ruzicka et al., 1993; Sano, 1994; Sano et al., 1993; Zhou et al., 1993).

We assessed our ability to detect the presence of an antigen by amplifying DNA attached to the antibody specific to that antigen. This method was found to be 100,000 times more sensitive than standard radioimmune assays. In principle, its sensitivity is unlimited; we should be able to detect single molecules. The problem, however, becomes background. In any real system, the amplification system is not useful unless background can be lowered sufficiently; thus, we have spent the last couple of years dealing with the issue of background. It is apparent that several more interesting technological advances will emerge as the genome project and related endeavors continue.

REFERENCES

Grothues, D., C. R. Cantor, and C. L. Smith. 1993. PCR amplification of megabase DNA with tagged random primers. Nucleic Acids Research 21:1321-1322.

Kinzler, K. W., and B. Vogelstein. 1989. Whole genome PCR; application to the identification of sequences bound by gene regulatory proteins. Nucleic Acids Research 17:3645-3653.

Ludecke, H. J., B. Senger, U. Claussen, and B. Horsthemke. 1989. Cloning defined regions of the human genome by microdissection of banded chromosomes and enzymatic amplification. Nature 338:348-350.

Ruzicka, V., W. Marz, A. Russ, and W. Gross. 1993. Immuno-PCR with a commercially available avidin system. Science 260:698-699.

Sano, T. 1993. Immuno-PCR: An ultra-sensitive antigen detection system using antibody-DNA conjugates. Experimental Medicine 11:1497-1499.

Sano, T. 1994. Immuno-PCR: Concept, methods, and prospects. Cell Technology 13:77-80.

Sano, T., C. L. Smith, and C. R. Cantor. 1992. Immuno-PCR: A very sensitive antigen detection system using a DNA-antibody conjugate. Science 258:120-122.

Sano, T., C. L. Smith, and C. R. Cantor. 1993. Response to a technical comment by Ruzicka et al. entitled "Immuno-PCR using a commercially available avidin system." Science 260:698-699.

Zhang, L., X. Cui, K. Schmitt, R. Hubert, W. Navidi, and N. Arnheim. 1992. Whole genome amplification from a single cell: Implications for genetic analysis. Proceedings of the National Academy of Sciences (USA) 89:5847-5851.

Zhou, H., R. J. Fisher, and T. S. Papas. 1993. Universal immuno-PCR for ultra-sensitive target protein detection. Nucleic Acids Research 21:6038-6039.

Biomarkers and Occupational Health: Progress and Perspectives. 1995. Pp. 264-274

Epigenetic Biomarkers: Potentials and Limitations

James E. Trosko

The term "biomarkers" is rarely defined but is usually assumed to mean a measurable entity that is found in some biological material and can be used as a predictor of some future health-related outcome. Potentially, biomarkers could be indicators of exposures to physical, chemical, or biological disease-influencing agents. They could also be measures of a specific genetic, developmental, or sex-related susceptibility. In addition, they could also be indicators of early, intermediate, or clinically apparent disease states.

In the quest for biomarkers to predict potential induction of birth defects, cancer, reproductive and neurological dysfunctions, cataracts, or cardiovascular diseases, it is important to emphasize that no single agent "causes" these multi-factorial diseases (Trosko, 1992). For example, a person exposed to radiation, cigarette smoke, asbestos, human immunodeficiency virus (HIV), or polybrominated biphenyls may or may not actually develop a given disease despite the presence of a measurable biomarker of exposure. Other important factors would include whether the measured levels of the exposure indices are of biological significance or are the result of using high resolution technology, and any of several factors that could either synergize or antagonize the impact of the biomarker being measured.

Homeostasis: The Ultimate Process to Maintain the State of Health

While the cell has been considered to be the ultimate unit of life, in a multi-cellular organism such as the human being, this concept has little functional meaning since the human being is more than simply a collection of 10^{14} cells. Human beings are "more than the sum of our parts"; we are a hierarchial system of cybernetically interacting processes (Brody, 1973; Potter, 1974). Although humans start from a single cell that contains the total genetic blueprint, the resulting organism encompasses a wide variety of cell, tissue, organ, system, and functional specificity. That one cell, the fertilized egg, by cell proliferation, differentiation, development, and adaptive responses of the differentiated cells, gives rise to an entire human being depending on a unique set of complex genetic and environmental interactions.

In principle, the delicate regulation of gene expression, cell proliferation, differentiation, and adaptive responses of the differentiated cells within and between tissues and organs is referred to as homeostasis. In the multi-cellular organism, *syncytia* of cells are the units of life. As shown in Figure 1, maintenance of "health" involves an integrated and tightly coordinated series of cellular communication processes, involving extra-, intra-, and gap junctional inter-communication (Trosko et al., 1994) within and between these syncytial units.

In effect, a perturbation of any of these interacting communication processes, by genetic or environmental factors or by endogenous or exogenous agents, will lead to perturbations in the other communication processes. Each perturbation will bring about consequences at the molecular, biochemical, cellular, physiological, or whole organismic functional levels. Some of these consequences could serve as given biomarkers.

Disruptors of Homeostasis: Adaptive Response or Disease-Causing?

In principle, disruption of the homeostatic control of fundamental biological processes, such as for cell proliferation, differentiation, development, tissue replacement/repair, and adaptive functions within and between differentiated cells and tissues, is necessary both for normal, healthy growth and for adaptive survival responses. It is the "unscheduled" acute or chronic disruption of homeostasis, via the perturbations of the intercellular communication mechanisms that can, and does, lead to various disease states.

These disruptions can be brought about by irreversible changes in the genetic material, either through hereditary or somatic mutations (mutagenesis), by cell death (cytotoxicity), or by potentially reversible changes in the expression of genes (epigenesis) (Trosko et al., 1993). By definition, mutations are qualitative or quantitative alterations in the genetic information in a cell. An epigenetic event is an alteration of the expression of the genetic information at the transcriptional, translational, or post-translational level (Trosko et al., 1990 a,b).

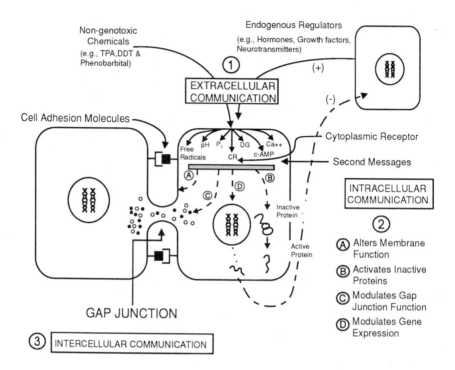

FIGURE 1. This scheme characterizes the postulated link between extracellular communication and intercellular communication via various intracellular transmembrane signaling mechanisms. It provides an integrating view of how the neuroendocrine-immune system (mind or brain/body connection) and other multisystem coordinations could occur. Although not shown here, activation or altered expression of various oncogenes also could contribute to the regulation of gap junction function.

The identification and use of "biomarkers" of exposure for particular disease states will be theoretically, technically, financially, ethically, and legally complex. For biomarkers of susceptibility, these issues will be even more problematic. Currently, there are a few biomarkers that could provide information about the potential health status of children (e.g., blood lead levels). However, the prevailing paradigm guiding much of the current research to find biomarkers for exposures to chemicals that are known or suspected of contributing to cancers, is that of DNA lesions and gene/chromosomal mutations. Clearly, this paradigm has been fueled by the idea that "carcinogens are mutagens" (Ames et al., 1973). While there is no doubt that mutations play a role in the carcinogenic process, either via genetic predispositions or by somatic mutations (Trosko et al., 1985), cancer, as well as other chronic diseases, is more than mutagenesis (Trosko and Chang, 1988 a,b). Therefore, the search for biomarkers of mutagen exposure is in fact justified to prevent hereditary and somatic mutations that could eventually lead to various diseases.

There are many difficulties in this task. For example, agents that induce mutations in cells can do so either by forming adducts or physical/chemical lesions in DNA or by altering the fidelity of DNA replication/repair of lesions. Agents that might lead to alterations in ploidy levels or to aneuploidy might not even interact directly with the DNA. Complicating the search for potential mutagens are the assays to measure mutation induction (Trosko, 1988). Whether the assay involves the detection of "lesions" in DNA (alkaline elusion for single strand breaks, ^{32}P-postlabelling of DNA, unscheduled DNA synthesis, comet formation, etc.) or chromosome aberrations (exchanges, aneuploidy, polyploidy, sister chromatid exchanges, etc.), all are subject to potential limitations and artifacts leading to misinterpretation of the "positive" results (Trosko, 1984).

Not all adducts/lesions in DNA lead to mutations, not all mutations lead to detectable diseases, and not all mutated disease-related genes are found in cells that can influence disease. A recent finding illustrates this point. The rodent carcinogen, methyl-nitrosourea (MNU), has been shown to be a mutagen in various in vitro assays. When administered to mice, tumors are formed and various mutations are observed in the oncogenes and tumor suppressor genes. However, Cha et al. (1994), in a study to detect potential mutations in spontaneous and MNU-induced liver tumors of mice, found that MNU seemed to select out and clonally expand the spontaneously existing mutated oncogene-containing cells. Thus, MNU appears to act more as a mitogen than a mutagen. One interpretation of the data is that MNU acts as an epigenetic promoter of pre-existing spontaneously mutated cells.

An important concept in the search for DNA lesion biomarkers is that cancer, as well as atherogenesis, is considered to be, for the most part, clonally derived from a single altered cell (Nowell, 1976; Wainscoat and Fey, 1990; Benditt and Benditt, 1973). If these cells are stem cell-derived, given the stem cell theory of carcinogenesis (Oberling, 1994; Potter, 1978; Sell, 1993), the task of determining the biomarkers of mutagen exposure would be almost impossible without some way to identify and measure the biomarkers in these few cells. Assessment of DNA lesions/repair is currently accomplished by analyzing DNA from whole tissues or populations of cells *in vitro*. Most of these cells are either differentiated or progenitor cells. Few are stem cells. If stem cells are undifferentiated, as by definition, then they probably have no or few metabolizing characteristics of their differentiated daughters. DNA repair capacities might also be different.

Although the search for biomarkers of cytotoxic events might seem easier, recent evidence indicates that there are two distinct mechanisms of cell death, namely, necrosis and apoptosis (Wyllie, 1992). Any biomarker of cell death (e.g., LDH release) would normally be only transient and unable to distinguish between those factors inducing necrosis from those inducing apoptosis. Agents inducing apoptosis might do so in a manner affecting only a few cells. As a matter of fact, a decrease in apoptosis might be a predictor of a potential disease in progress. Recently, it has been shown that several tumor-promoting chemicals (Bursch et al., 1992) and several oncogenes (McDonnell, 1993) block apoptosis, while anti-tumor promoters (Gelia et al., 1993; Alles and Sulik, 1989) and a major tumor suppressor gene (Lane, 1993) enhances apoptosis. This has led to the hypothesis that epigenetic agents that block gap junctional intercellular communication can block apoptosis (Trosko and Goodman, 1994).

The search for biomarkers of diseases involving epigenetic mechanisms will also be difficult because controlled epigenetic events are occurring constantly in a healthy body. Alterations in gene expression at the transcriptional level have been associated with both normal cellular development and differentiation and cell cycle regulation, as well as with carcinogenesis and genetic imprinting (Holliday, 1987; Hoffman, 1984; Li et al., 1993).

Many physical and chemical agents known to be associated with the induction of birth defects (i.e., thalidomide), cancer, neurological and reproductive diseases (e.g., asbestos, dioxins, polybrominated biphenyls, reserpine, phorbol esters) can alter gene expression at the transcriptional, translational, or post-translational levels (Trosko and Chang, 1988 a,b). Even physical agents, such as ionizing and ultraviolet light radiations, have been shown to have epigenetic effects as well as genetic effects (Warmuth et al., 1994; Kharbunda et al., 1994; Boothman and Lee, 1991).

The mechanism by which epigenetic agents can disrupt homeostasis is by interfering with the elaborate and tightly coordinated intercellular communication (IC) processes which involve extra-, intra-, and inter-cellular communication (GJIC). Disruption can occur by up-or-down regulation of extra-cellular communicating signals (e.g., hormones, growth factors, neurotransmitters), which in turn trigger intra-cellular signals (e.g., calcium, c-AMP, pH level, oxygen radical species, protein kinases, etc.). These signals, in turn, can up- or down-regulate GJIC (Trosko et al., 1990 a,b). Unfortunately, the cell cannot distinguish between extra-cellular signals from exogenous sources (e.g., pollutants, medication, food additives, etc.) and those from endogenous sources (e.g., diurnal hormone signals, normal growth, or wound-healing factors, etc.).

The biological consequence of this cascade is that the cell's physiological state changes so that it can either proliferate, differentiate, or adaptively respond. In the mature organism, the vast majority of the cells, both in solid and soft tissues, are in the G_o or resting state. In solid tissues, contact inhibition, mediated by GJIC, maintains the cells in this quiescent state. Modulation of GJIC by extra- and intra-cellular communication signals can either block the transfer of mitogenic suppressing signals. Disruption of GJIC can have adaptive and maladaptive consequences, ranging from normal cell growth, differentiation and wound-healing to teratogenesis, tumor promotion, reproductive dysfunction, and neurological disorders (Trosko et al., 1993).

Recently, markers of cell proliferation have been suggested as potential biomarkers of carcinogenesis and tumor promoters (Ames and Gold, 1990; Cohen and Ellwein, 1990). Unfortunately, although cell proliferation can be an indicator of compensatory hyperplasia due to cytotoxicity and can be responsible, in part, for clonal expansion of initiated cells during tumor promotion, some tumor promoters actually suppress cell proliferation in normal cells, stimulate only initiated cells to proliferate, or block apoptosis, thus preventing the death of initiated cells (Schulte-Hermann et al., 1990). Markers of general mitogenesis would probably not prove to be a reliable biomarker (Melnick and Huff, 1993; Ward et al., 1993).

Epigenetic Biomarkers: Future Considerations

While the search for biomarkers of genotoxic lesions, of susceptibilities to genotoxic agents, and of disease indicators having mutational origins should continue, the search for biomarkers of epigenetic "lesions," susceptibilities to epigenetic agents, and diseases having an epigenetic origin will be extraordinarily difficult. Genotoxic agents do have some potential for leaving a "molecular fingerprint," either as a persistent lesion in DNA or as a unique pattern of mutations induced in the DNA. Molecular epidemiology has its foundation based on these assumptions (Shields and Harris, 1991). In principal, environmental epigenetic inducers of disease will not leave physical adducts in DNA or alterations in genetic information. Abnormal expression of certain gene products might be detectable. In some cases of an epigenetic agent-induction of disease states (e.g., thalidomide- or retinoid-induced teratogenesis), while the molecular/biochemical mechanisms might be reversible, the cellular effect during a critical state of embryonic development is irreversible.

There is a real need to search for biomarkers for epigenetic agents. Several environment toxicants known to "cause" cancer, neurological effects or reproductive disorders are, in fact, not genotoxic (Trosko and Chang, 1988). Dioxins, polychlorinated and polybrominated biphenyls, DDT, dieldrin, aldrin, toxaphene, mirex, kepone, etc., are clearly acting via epigenetic mechanisms. Almost all of the effort in risk assessment is based on the identification of potential genotoxicity of known or suspected environmental pollutants. Little or no emphasis is placed on the fact that many of these known pollutants are not genotoxic and little effort is placed in putting considerations of epigenetic mechanisms into risk assessment analyses (Trosko et al., 1994).

Given the current theoretical and practical limitations in finding non-invasive, valid *in vivo* biomarkers for epigenetic toxicants, it would appear that both basic and practical studies in rodent and human *in vitro* systems will be necessary to identify potential epigenetic agents and to understand their mechanisms of action. The inherent limitation in all *in vitro* studies is that they can never perfectly mimic the *in vivo* situation. However, gap junctional intercellular communication does exist both *in vitro* and *in vivo* and since GJIC is the common endpoint for epigenetic agents that affect transcription, translation, or post-translational levels by a diverse series of mechanisms, GJIC could be used *in vitro* to determine if a given potential toxicant can: (a) modulate GJIC in rodent or human cells; (b) modulate specific cells of particular tissues in either species; (c) act via one or another molecular or biochemical mechanism; (d) be modified by mixtures of interacting chemicals; and/or (e) have no effect or threshold levels of action.

Finally, as a further complication in the search for critical biomarkers of exposure, susceptibility or disease precursors, several diseases have a clonal origin from a single dysfunctional stem cell. To find non-invasive strategies for identifying these few cells will be extremely difficult.

ACKNOWLEDGEMENTS

The research on which this analysis was based has been supported by grants from the U.S. Air Force Office of Scientific Research (F49620-92-J-0293), NCI (Ca 21104), NIEHS (2-P42ES04911), Michigan Great Lakes Protection Fund, and a gift from the Acrylonitrile Group. In addition the author wishes to acknowledge the excellent word processing skills of Mrs. Robbyn Davenport.

REFERENCES

Alles, A. J., and K. Sulik. 1989. Retinoic acid-induced limb-reduction defects perturbations of zones of programmed cell death as a pathogenetic mechanism. Teratology 40:163-171.

Ames, B. N., W. E. Dunston, E. Yamasaki, and F. D. Lee. 1973. Carcinogens are mutagens: A simple test system combining liver homogenates for activation and bacteria for detection. Proceedings of the National Academy of Sciences (USA) 70:2281-2285.

Ames, B. N., and L. S. Gold. 1990. Too many rodent carcinogens: Mitogenesis increases mutagenesis. Science 249:970-971.

Benditt, E. P., and J. M. Benditt. 1973. Evidence for a monoclonal origin of human atherosclerotic plaques. Proceedings of the National Academy of Sciences (USA) 70:1733-1736.

Boothman, D. A., and S. W. Lee. 1991. Regulation of gene expression in mammalian cells following ionizing radiation. Yokohama Medical Bulletin 42:137-149.

Brody, H. 1973. A systems view of man: Implications for medicine, science and ethics. Perspectives in Biology and Medicine 7:71-92.

Bursch, W., F. Oberhammer, and R. Schulte-Hermann. 1992. Cell death by apoptosis and its protective role against disease. Trends in Pharmacological Sciences 13:245-251.

Cha, R. S., W. G. Thilly, and H. Zarbl. 1994. N-Nitroso-N-methylurea-induced rat mammary tumors arise from cells with pre-existing oncogenic Hras I gene mutations. Proceedings of the National Academy of Sciences (USA) 91:3749-3753.

Cohen, S. M., and L. B. Ellwein. 1990. Cell proliferation in carcinogenesis. Science 249:1007-1011.

Gelia, D., A. Aiello, L. Lombardi, P. G. Pelicci, F. Grignani, F. Formelli, S. Menard, A. Costa, U. Veronesi, and M. A. Pierotti. 1993. N-(4-Hydroxyphenyl) retinamide induces apoptosis of malignant hemopoietic cell lines including those unresponsive to retinoic acid. Cancer Research 53:6036-6041.

Hoffman, R. M. 1984. Altered methionine metabolism, DNA methylation and oncogene expression in carcinogenesis. Biochim. Biophys. Acta 738:49-87.

Holliday, R. 1987. The inheritance of epigenetic defects. Science 1238: 163-170.

Kharbunda, S., A. Saleem, R. Datta, Z. M. Yuan, R. Weichselbaum, and D. Kufe. 1994. Ionizing radiation induces rapid tyrosine phosphorylation of $_p34^{cdc2}$. Cancer Research 54:1412-1414.

Lane, D. P. 1993. A death in the life of P^{53}. Nature 362:786-787.

Li, E., C. Beard, and R. Jaenisch. 1993. Role of DNA methylation in genomic imprinting. Nature. 366:362-365.

McDonnell, T. J. 1993. Cell division versus cell death: A functional model of multistep neoplasia. Molecular Carcinogenesis 8:209-213.

Melnick, R. L., and J. Huff. 1993. Liver carcinogenesis is not a predicted outcome of chemically induced hepatocyte proliferation. Toxicology and Industrial Health 9:415-438.

Nowell, P. C. 1976. The clonal evolution of tumor cell population. Science 194:23-28.

Oberling, C. 1944. The riddle of cancer. Yale University Press: New Haven, Connecticut.

Potter, V. R. 1974. Probabilistic aspects of the human cybernetic machine. Perspectives in Biology and Medicine 17:164-183.

Potter, V. R. 1978. Phenotypic diversity in experimental hepatomas: The concept of partially blocked ontogeny. British Journal of Cancer 38:1-23.

Schulte-Hermann, R., I. Timmermann-Trosiener, G. Barthel, and W. Bursch. 1990. DNA synthesis, apoptosis and phenotypic expression as determinants of growth of altered foci in rat liver during phenobarbital promotion. Cancer Research 50:5127-5135.

Sell, S. 1993. Cellular origin of cancer: Dedifferentiation or stem cell maturation arrest? Environmental Health Perspectives 5:15-26.

Shields, P. G., and C. C. Harris. 1991. Molecular epidemiology and the genetics of environmental cancer. Journal of the American Medical Association 266:681-687.

Trosko, J. E. 1984. A new paradigm is needed in toxicological evaluation. Environmental Mutagenesis 6:767-769.

Trosko, J. E. 1988. A failed paradigm: Carcinogenesis is more than mutagenesis. Mutagenesis 3:363-366.

Trosko, J. E. 1992. Does radiation cause cancer? RERF Update 4:3-5.

Trosko, J. E. and C. C. Chang. 1988a. The role of inhibited intercellular communication in carcinogenesis: Implications for risk assessment from exposure to chemicals. Pp. 165-179 in Biologically Based Methods for Cancer Risk Assessment. C. C. Travis, ed. Plenum Press: New York.

Trosko, J. E., and C. C. Chang. 1988b. Nongenotoxic mechanisms in carcinogenesis: Role of inhibited intercellular communication. Pp. 139-170 in Banbury Report 31: Carcinogen Risk Assessment: New Directions in the Qualitative and Quantitative Aspects. R. W. Hart and F. G. Hoerger, eds. Cold Spring Harbor Laboratory Press: Cold Spring Harbor, New York.

Trosko, J. E., and J. I. Goodman. 1994. Intercellular communication may facilitate apoptosis: Implications for tumor promotion. Molecular Carcinogenesis (in press).

Trosko, J. E., V. M. Riccardi, C. C. Chang, S. Warren, and M. Wade. 1985. Genetic predispositions to initiation or promotion phases in human carcinogenesis. Pp. 13-37 in Biomarkers, Genetics and Cancers. H. Anton-Guirgis and H. T. Lynch, eds. Van Nostrand Rheinhold, Inc.: New York.

Trosko, J. E., C. C. Chang, and B. V. Madhukar. 1990a. *In vitro* analysis of modulators of intercellular communication: Implications for biologically based risk assessment models for chemical exposure. Toxicology *In Vitro* 4:635-643.

Trosko, J. E., C. C. Chang, B. V. Madhukar, and J. E. Klaunig. 1990b. Chemical, oncogene and growth factor inhibition of gap junction intercellular communication: An integrative hypothesis of carcinogenesis. Pathobiology 58:265-278.

Trosko, J. E., B. V. Madhukar, and C. C. Chang. 1993. Endogenous and exogenous modulation of gap junctional intercellular communication: Toxicological and pharmacological implications. Life Sciences 53:1-19.

Trosko, J. E., C.C. Chang, and B. V. Madhukar. 1994. The role of modulated gap junctional intercellular communication in epigenetic toxicology. Risk Analysis. 14:303-312.

Wainscoat, J. S., and M. F. Fey. 1990. Assessment of clonality in human tumors. A review. Cancer Research 50:1355-1360.

Ward, J. M., H. Uno, Y. Kurata, C. M. Weghorst, and J. J. Jang. 1993. Cell proliferation not associated with carcinogenesis in rodents and humans. Environmental Health Perspectives 10:125-136.

Warmuth, I., Y. Harth, M. S. Matsui, N. Wang, and V. A. DeLeo. 1994. Ultraviolet radiation induces phosphorylation of the epidermal growth factor receptor. Cancer Research 54:374-376.

Wyllie, A. H. 1992. Apoptosis and the regulation of cell numbers in normal and neoplastic tissues: An overview. Cancer Metastasis Reviews 11:95-103.

Biomarkers and Occupational Health: Progress and Perspectives. 1995. Pp. 275-289

Flow Cytometry: A Powerful Technology for Measuring Biomarkers

James H. Jett

Biomarkers can be broadly defined as measurable characteristics of a biological system that can change upon exposure to a physical or chemical insult. While this definition can be further refined, it is sufficient for the purposes of demonstrating the advantages of flow cytometry for use in providing quantitative measurements of biomarkers. Flow cytometry and cell sorting technologies have emerged during the past 25 years to take their place alongside other essential tools used in biology such as optical and electron microscopy. This paper describes the basics of flow cytometry technology, and provides examples of applications of this technology in the field of biomarkers. The examples of the uses of flow cytometry for biomarker quantification presented in this paper are meant to be illustrative of the potential of this technology rather than to provide an exhaustive review of this technical area.

Flow Cytometry

The term "moving microscopy" is useful for describing how a flow cytometer works. Flow cytometers make many of the same measurements that are made by modern digital microscopes (Givan, 1992; Shapiro, 1988; Watson, 1991). However, there are important differences between the two approaches that result in differences in measurement resolution and speed. Flow cytometry has its roots in the development of Coulter volume measurements of cell size. Coulter volume measurements are made by passing cells through a small orifice that is several times larger than a cell. By attaching electrodes on both sides of the orifice to a constant voltage source, a small electrical current passes through the opening and can be measured. When a cell passes through this opening, the resistance increases and the current decreases. The change in current is measured and is directly related to the volume of the cell. Cells are transported through the orifice one at a time by a fluid transport system.

Because cell volume measurements are not distinctive enough to discriminate between cell types, new approaches to rapid quantification of cellular properties were developed. The most successful of these developments was based on hydrodynamic focusing in which a line of flowing cells is transported through a sensor region. Figure 1 schematically demonstrates the principle. The first step is to prepare a single cell suspension of stained cells. The stains are usually fluorescent reagents that bind specifically and stoichiometrically to a cellular component that is to be measured. As will be discussed later, it is possible to stain cells simultaneously with multiple reagents. The cells are introduced into a sheath flow cuvette through a sample inlet tube that is surrounded by a flowing sheath fluid. As illustrated in Figure 1, the combination of the internal geometry of the flow cuvette and the proper ratio of the sample and sheath flow rates hydrodynamically focuses the sample stream down to a diameter that is on the order of 10 microns, or a typical cell diameter. In addition to aligning the cells so that they are like beads on a string, the hydrodynamic focusing also accelerates the sample stream to a velocity that is measured in meters per second.

Once the cell stream is hydrodynamically focused, it passes through one or multiple laser beams. The laser light is focused onto the sample stream to provide intense illumination of the cells. Under typical cytometer operating conditions, the laser beam spans 50 microns in the direction of the flow and the sample stream has a velocity of 10 meters per second which results in the cells being illuminated by the laser beam for 5 microseconds. This short measuring time is the basis for the first characterizing property of flow cytometers which is the ability to make measurements on a large number of cells in a short amount of time. Thus, accurate measurements of distributions of cellular properties can be made rapidly.

When a cell passes through the laser beam, the fluorescent reagent bound to the cell is excited and emits fluorescence that is collected optically and directed to a photodetector. Optical filters separate the scattered laser light from the fluorescence before the collected light reaches the detector. The signal produced by the detector is amplified, processed, digitized, and the results stored in a computer. Under the proper experimental conditions, analysis of the fluorescence signal yields information about the width of the particle passing through the laser beam by measurement of the signal width, the maximum staining intensity by measuring the amplitude of the signal, and the total staining intensity by measuring the integral of the signal. If the fluorophor binds to DNA, the nuclear width, maximum DNA density, and total DNA content are determined. If the fluorophor binds to protein, analogous values are measured for the cell as a whole.

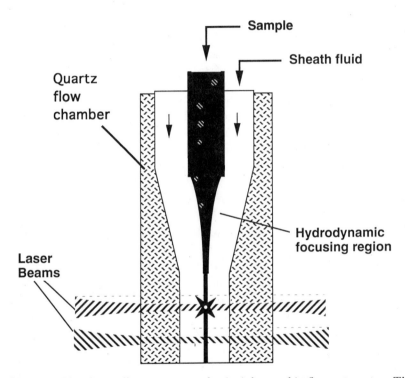

FIGURE 1. This figure illustrates several principles used in flow cytometry. The sample is introduced at the top of the device as a suspension of particles: cells, chromosomes, or molecules. The sample fluid is forced out of the sample inlet tube into the surrounding sheath fluid that is flowing from top to bottom. As the sample enters the sheath fluid, it is hydrodynamically focused to produce a fine sample stream that is approximately one cell diameter in width and flows at a velocity of several meters per second. One or more laser beams intercept the focused sample stream and excite fluorophors bound to the sample constituents. Signals from the emitted fluorescence are directed to photo detectors, digitized, and stored in a computer for post-acquisition analysis.

In special configurations of the apparatus, other properties of the fluorescence emissions are measured. For example, analysis of the state of polarization of the fluorescence is indicative of the local environment of the fluorophor and has been used in studies of membrane fluidity. For the calcium-indicating dye Indo-1, the ratio of the emission intensities in two wavelength regions provides a measure of the calcium ion concentration in the cells.

All particles, whether stained or not, scatter the illumination light as they pass through it. Measurement of the scattered light in the forward direction yields information about the size of the particles. The intensity of the light scattered at 90 degrees from the direction of illumination is related to a complex convolution of the size of the particle, its surface texture, and its internal structures such as lipid vesicles. While these statements are generally true, light scatter measurements are very complex and must be carefully interpreted (Rauchman et al., 1992; Salzman et al., 1990). All commercial flow cytometers are capable of measuring forward angle and right angle light scatter. Such measurements are used primarily to distinguish subpopulations of cells in a sample.

Fluorescent reagents used in flow cytometry can be broadly categorized into those that bind to DNA, those that measure physiological parameters such as pH, and those that measure intra-cellular and cell surface antigens. The development of monoclonal antibody technology has had a major impact on flow cytometry by providing reagents that uniquely label a huge variety of molecules within and on the surface of cells. Monoclonal antibodies are valuable tools in the analysis of immunological systems as well as in the analysis of intracellular antigens.

The advantages of flow cytometry over conventional fluorescence microscopy are related to the speed of data collection. A typical commercial flow cytometer is capable of analyzing cells at rates up to 10,000 cells per second. High analysis rates facilitate the collection of large amounts of data per sample analyzed. It is not unusual to collect data on 5,000,000 cells in a sample. The large amounts of data provide information on the distribution of the measured parameters with high statistical precision. Biomarker assays in which the signal is due to a rare event, such as a mutation assay, must rely on the ability to screen large numbers of cells to obtain statistically meaningful data.

There is a price to pay for the high rates of data collection. Flow cytometric measurements are often referred to as zero resolution measurements. That is, no morphological information is obtained during the measurement, only the total amount of a cellular component is determined. As indicated above, under appropriate conditions, it is possible to obtain one-dimensional information in a flow measurement. At least one research system (Wheeless, 1990) records three orthogonal one-dimensional images of squamous cells to detect cancer cells. In contrast, modern microscope systems acquire two- and three-dimensional cellular images with exquisite resolution.

Flow cytometric measurements are being made on at least five levels of biological complexity. The highest level of particle complexity is multicellular systems such as multicellular spheroid models of tumors (Freyer et al., 1987). Analysis at the cellular level is the most obvious field of application of flow cytometry as well as the most exploited. Cellular analysis includes not only the analysis of mammalian cells but also the analysis of bacterial cells and virus particles. Chromosomes are essentially the only isolated subcellular particles that have been extensively analyzed by flow cytometry. Flow karyotyping is used to study normal chromosomal complements, to determine alterations in chromosomes, and as a method of identification for chromosome sorting. Large numbers of chromosomes have been, and are being, sorted to provide chromosome-specific template material for the construction of recombinant DNA libraries (VanDilla et al., 1986).

In one sense, all flow cytometric measurements that use fluorescent reagents are molecular measurements. Dyes that bind specifically to a cellular component provide a means of measuring the levels of that molecule in a cell. Recently, flow cytometric measurements of molecules in solution have been made. A fluorescence-based homogeneous immunoassay has been developed (Saunders et al., 1985) which is capable of determining the concentrations of antigen molecules in solution. Measurements at the submolecular level elicit information about portions of molecules. For example, the presence of a specific configuration of an amino acid in a protein can be determined by antibody labeling, or a specific DNA sequence in a fragment of DNA can be measured by labeled oligonucleotide hybridization.

APPLICATIONS OF FLOW CYTOMETRY TO BIOMARKER MEASUREMENTS

Measurements of biomarkers by flow cytometry are extensive and cover many aspects of cellular biology (Dallas and Evans, 1990). There are at least five general categories of flow cytometric measurements that have been applied to biomarkers. They are DNA alterations, physiological measurements, histopathological analyses, immunological measurements, and molecular marker determinations. Specific examples of biomarker measurements by flow cytometry will be discussed below.

DNA Content Analysis

Cellular DNA content measurements were among the first measurement techniques developed and consequently the first to be applied to biomarker measurements. Cellular DNA content measurements are the most simple flow cytometry measurements to make on single cell suspensions. There is a wide variety of dyes that can be used singly, or in combination with other staining reagents (Darzynkiewicz and Crissman, 1990), to label cellular DNA. Diploid cellular DNA content is easily measured with a precision better than 1.0% which provides a means of identifying the presence of subpopulations of cells with aneuploid DNA content. Bickham and coworkers (Bickman et al., 1992; Bickman et al., 1994) have studied the effects of triethylenemelamine on somatic and testicular tissues of the rat. Their analyses relied upon quantification of changes induced in the coefficient of variation of the G1 peak in DNA distributions in the treated populations. In some cases, they were able to follow recovery of the animal after exposure to the mutagen by monitoring DNA distribution statistics.

Cell Cycle Analysis

In addition to ploidy measurements, cellular DNA content measurements are used in studies of cell cycle distributions. It is well known that there are a number of physical and chemical agents that affect cell cycle traverse. The transient G2 block induced by radiation exposure is one of the first cell cycle perturbations studied. Along with the cell preparation and analysis techniques, algorithms have been developed that determine cell cycle distribution parameters from the measured DNA content distributions. Maier and Schwalder (1986) have analyzed cellular DNA and protein content distributions for cultured rat fibroblasts after exposure to a variety of chemotherapeutic agents. They analyzed alterations in the measured distributions to collect information about cytostatic effects of the drugs tested.

New methods of studying the kinetics of cell growth have emerged in the past few years. For example, Crissman and Steinkamp (1987) perfected the use of BrdU incorporation to track cells around the cell cycle. Their technique, which does not use antibody labeling of the incorporated BrdU, uses two DNA binding dyes. The fluorescence of one (mithramycin) is not quenched by BrdU incorporation and the fluorescence of the other dye (Hoescht 33342) is quenched. Thus, the fluorescence from mithramycin provides information on the total DNA content in each cell and the difference between the two signals is related to the BrdU incorporation. With this technique, they have been able to follow cells in perturbed populations around several cycles. BrdU incorporation measurements have been used to analyze early changes in cell cycle kinetics after low dose irradiation (Rhein et al., 1994).

Subcellular DNA Measurements

DNA content measurements are used to determine the induction of micronucleus formation in mammalian cell cultures after radiation exposure. Schreiber et al. (1992) developed a technique for the rapid quantification of micronucleus formation which detects micronuclei with a DNA content above 0.5% of the DNA contained in a G1 nucleus. They clearly demonstrated the efficacy of this approach for determining the dose response of micronucleus formation due to ionizing radiation. Another example of a DNA content measurement at the subcellular level is flow karyotype analysis. Over the years, techniques have been perfected for isolating and staining chromosomes from many cell types and from numerous animal species. The effects of radiation on the karyotype of dividing human cells (Fantes et al., 1983) and Chinese hamster ovary cells (Welleweerd et al., 1984) have been studied by several groups using flow cytometry. Some studies have revealed dose-dependent karyotype alterations and correlations with cell survival.

Sperm Maturation

Spermatogenesis is a complex multi-stage process that is sensitive to environmental factors. Evenson et al. (1986) have applied flow cytometry to the study of spermatogenesis and male fertility. They have been able to identify several stages during sperm maturation by a variety of measurements, some of which provide DNA conformational information. A sperm viability assay measures DNA staining by a dye that does not permeate an intact membrane in conjunction with a mitochondrial function indicator.

Immunological Measurements

A recent National Research Council study (1992) lists five areas in which immunologically-based measurements are made. Flow cytometry has been applied to measurements in each of these areas: total levels of antibodies, total levels of immunoglobin, absolute numbers of lymphocytes, relative numbers of lymphocytes (immunophenotyping), and reactions to antigens/mitogens.

For example, the relative number of lymphocytes in an organism can be quickly quantified with the use of monoclonal antibodies that identify not only T-lymphocytes and B-lymphocytes but also subpopulations within each lineage (Darzynkiewicz and Crissman, 1990; Laerum and Bjerknes, 1992). This type of analysis is performed on a routine basis in many hospitals throughout the world. The assay is used to assess immune state and as a diagnostic and prognostic tool in leukemia and lymphoma cases. Immunophenotyping is an example of an assay that can fully exploit multiple labeling reagents and multiple color detection. Three-color lymphocyte analysis is a routine analysis procedure in clinical laboratories. Immunophenotyping has been applied in a variety of immunotoxicology studies (Hudson et al., 1985).

A flow cytometric fluoroimmunoassay that was developed by Saunders et al. (1985) of Los Alamos has exquisite sensitivity and is a homogeneous immunoassay. The assay can be used to quantify concentrations of any molecule for which an antibody can be produced. In the competitive binding mode, the detection limit of the assay for horseradish peroxidase (as the antigen) was 10^{-12} molar and the detection limit in the sandwich mode was 10^{-14} molar. Thus, this type of flow cytometric immunoassay can be used as the basis for the development of a soluble biomarker measurement system.

Mutation Assays and Rare Event Detection

At least three mutation assays (Kushiro et al., 1992; Kyoizumi et al., 1992; Langlois et al., 1990) have been developed which rely upon the rare event detection capability of flow cytometric measurements. As indicated above, these assays require collection of data from large numbers of cells, on the order of five million cells or more. The first flow cytometric mutation assay developed detected mutations in the human glycophorin A locus in humans due to radiation exposure. Using antibodies to detect the presence of MN blood group antigens on the surfaces of red blood cells, mutations rates on the order of 10^{-5} to 10^{-6} have been detected.

The three assays referenced above examine direct effects of the mutations by analyzing the presence or absence of specific molecules. It is also possible to measure indirect, or secondary, effects of a mutation. A mutation or a gene expression assay could be developed based upon analysis of membrane transport properties. Rauchman et al. (1992) have developed an assay for the analysis of glucose transport in cells transfected with a glucose permease gene. Since this is a functional assay, and not an assay for the presence or absence of a specific molecule, it could be used to quantify the secondary effects of a gene mutation by measuring alterations in the cell's transport capabilities.

Gene Expression Assays

Antibodies can be used as labels to quantify the products of gene expression. As an example of this, Johnson et al. (1993) have been able to measure the changes in the levels of the cell damage response protein, p53, in cells exposed to X-rays. Preliminary results indicate that a dose-dependent response is observed and that the exposure level for detecting the onset of the response is on the order of 0.01 Gray.

Cell Physiology Assays

In addition to the assays described to this point, there are several assays of cellular physiological parameters. Cellular sodium and calcium ion concentrations are measurable by flow cytometry using ratiometric dyes. Changes in membrane fluidity can be determined by measurement of changes in the polarization of fluorescence from the membrane bound dye. These techniques have been used to study the effects of heat on membrane properties (Dynlacht and Fox, 1992a, b).

RECENT DEVELOPMENTS AND NEW HORIZONS

A number of flow cytometric assays or analysis techniques that are currently available are not used to measure biomarkers. Enzyme activity in living cells can be measured by recording the time that a cell passes through the interrogation point (Martin and Swartzendruber, 1980). This type of assay uses a fluorogenic substrate for the enzyme such as the one designed by Marrone et al. (1991) for the p450 cholesterol side chain cleavage enzyme that makes up part of the steroid production cycle. Additional uses of the time parameter include monitoring of macromolecular assembly in cell activation processes.

There are several new flow cytometry instruments that can be used for cellular and subcellular analysis. With two of these new technologies it is possible to do single cell spectroscopy. A Fourier transform cytometer is capable of determining the emission spectrum on a cell by cell basis (Buican, 1989). This cytometer is potentially useful for studies with indicator dyes that change their emission spectrum in response to local environmental conditions. Single cell time spectroscopy in the nanosecond time range is possible by phase sensitive flow cytometry (Steinkamp and Crissman, 1993; Steinkamp et al., 1993). This technology allows the use of fluorescence decay time as a new parameter. It can be used to separate the signals from spectrally overlapping fluorophors that have differing decay times. Phase sensitive flow cytometry methods can also be applied to quantification of changes in fluorescence decay times due to environmental conditions.

Recent advances in flow cytometry include improvements in sensitivity to the point that it is possible to detect individual fluorescent molecules (Wilkerson et al., 1993) as they pass through the interrogation laser beam. This represents an improvement in detection sensitivity of 10,000-fold over a six-year period. Two biological applications of this ultrasensitive form of flow cytometry have emerged. As part of the Los Alamos Human Genome effort, a radically new approach to sequencing DNA, based upon single molecule detection by flow cytometry, is being developed (Ambrose et al., 1993). This new technique will be capable of sequencing large pieces (40 kilobases) of DNA at rates approaching several hundred bases per second. When this technique is available for general use, it will be applicable to the detection of specific mutations by direct sequencing of the region of interest.

DNA fragment size analysis is currently accomplished exclusively by gel electrophoresis in research laboratories worldwide. Analysis of DNA fragment size distributions by flow cytometry is 1,000-fold or more faster than existing techniques and requires significantly less material for analysis. This is accomplished by staining DNA fragments with bright fluorescent dyes that bind stoichiometrically and analyzing them individually in a flow cytometer (Goodwin et al., 1993). In contrast to electrophoresis, the response of the system to a fragment of DNA is linearly related to its size. The linearity of the response has been demonstrated over the size range of 1.5 to 145 kilobase pairs. Future refinements of the technique will add the ability to detect two colors of fluorescence which will enable not only measurement of fragment size but also the detention of a specific DNA sequence on the fragment by hybridization of fluorescent labeled probes.

CONCLUSIONS

Over the past 25 years, there has been a continuous evolution of flow cytometry technology. Future instrumental developments will result in the use of new laser light sources that operate further into the UV and IR regions of the spectrum. New applications of flow cytometry will be forthcoming as staining reagents are developed that take advantage of the new light sources. In the future, a flow cytometer on a chip will be available that integrates a solid state light source with the flow channel and detectors. This will bring the size and complexity of a system down to the point that a battery-operated instrument suitable for field operation is practical.

In summary, the applications of flow cytometry-based measurements of biomarkers, past and future, are numerous and, in some instances, provide information that could not otherwise be readily obtained. The multiparameter measurement capabilities of flow cytometry provide methods for dissecting cell populations with ever increasing resolution and for determining subtle changes in cellular characteristics. The ability to make high-speed measurements on large numbers of cells in short periods of time enables the quantification of rare cellular events. The range of applications is from the level of whole cells to the submolecular level. As new biomarkers are developed, flow cytometry will be an important and powerful technique for their quantification.

ACKNOWLEDGEMENTS

Several investigators have assisted in the development of the ideas incorporated into this paper. The assistance of N. Johnson, S. Burchiel, B. Marrone, R. Habbersett, J. Martin, and the responders to my inquiry on the Flow Cytometry bulletin board is gratefully acknowledged. This work is supported by the U.S. Department of Energy, Office of Health and Environmental Research and the National Flow Cytometry Resource (NIH NCRR Grant RR01315).

REFERENCES

Ambrose, W. P., P. M. Goodwin, J. H. Jett, M. E. Johnson, J. C. Martin, B. L. Marrone, J. A. Schecker, C. W. Wilkerson, R. A. Keller, A. Haces, P. Shih, and J. D. Harding. 1993. Application of single molecule detection to DNA sequencing and sizing. Ber. Bundsenges. Phys. Chem. 97:1535-1542.

Bickman, J. W., V. L. Sawin, D. W. Burton, and K. McBee. 1992. Flow-cytometric analysis of the effects of triethylenemelamine of somatic and testicular tissues of the rat. Cytometry 13:368-373.

Bickman, J. W., V. L. Sawin, K. McBee, M. J. Smolen, and J. N. Derr. 1994. Further flow cytometric studies of the effects of triethylenemelamine of somatic and testicular tissues of the rat. Cytometry 15:222-229.

Biologic Markers in Immunotoxicology. 1992. Committee on Biologic Markers. Washington, D.C.: National Academy Press.

Buican, T. N. 1989. Real-time transform spectroscopy for fluorescence imaging and flow cytometry. Proceedings of the SPIE 1205:103-112.

Crissman, H. A., and J. A. Steinkamp. 1987. A new method for rapid and sensitive detection of bromodeoxyuridine in DNA-replicating cells. Experimental Cell Research 173:256-261.

Dallas, C. E., and D. L. Evans. 1990. Flow cytometry in toxicology analysis. Nature 345:557-558.

Darzynkiewicz, Z. and Crissman, H. A., eds. 1990. Flow Cytometry. Methods in Cell Biology. Volume 33. San Diego, California. Academic Press.

Dynlacht, J. R., and M. H. Fox. 1992a. The effect of $45^{o}C$ hyperthermia of the membrane fluidity of cells of several lines. Radiation Research 130:55-60.

Dynlatch, J. R., and M. H. Fox. 1992b. Heat-induced changes in the membrane fluidity of Chinese hamster ovary cells measured by flow cytometry. Radiation Research 130:48-54.

Evenson, D., Z. Darzynkiewicz, L. Jost, F. Janca, and B. Ballachey. 1986. Changes in accessibility of DNA to various fluorochromes during spermatogenesis. Cytometry 7:45-53.

Fantes, J. A., D. K. Green, J. K. Elder, P. Malloy, and H. J. Evans. 1983. Detection radiation damage to human chromosomes by flow cytometry. Mutation Research 119:161-168.

Freyer, J. P., M. E. Wilder, and J. H. Jett. 1987. Viable sorting of intact multicellular spheroids by flow cytometry. Cytometry 8:427-436.

Givan, A. L. 1992. Flow Cytometry: First Principles. New York, NY: Wiley-Liss.

Goodwin, P. M., M. E. Johnson, J. C. Martin, W. P. Ambrose, B. L. Marrone, J. H. Jett, and R. A. Keller. 1993. Rapid sizing of individual fluorescently stained DNA fragments by flow cytometry. Nucleic Acids Research 21:803-806.

Hudson, J. L., R. E. Duque, and E. J. Lovett III. 1985. Applications of flow cytometry in immunotoxicology. Pp. 159-177 in: Immunotoxicology and Immunopharmacology. J. H. Dean, M. I. Luster, A. E. Munson, and H. Amos, eds. New York: Raven Press.

Johnson, N. F., R. J. Jaramillo, and A. W. Hickman. 1993. Wild-type p53 expression in cultured lung epithelial cells exposed to alpha particles. Inhalation Toxicology Research Institute Annual Report.

Kushiro, J., Y. Hirai, Y. Kusunoki, S. Kyoizumi, Y. Kodama, A. Wakisaka, A. Jeffreys, J. Cologne, K. Dohi, N. Nakamura, and M. Akiyama. 1992. Development of a flow-cytometric HLA-A locus mutation assay for human peripheral blood lymphocytes. Mutation Research 272:17-29.

Kyoizumi, S., S. Umeki, M. Akiyama, Y. Hirai, Y. Kusunoki, N. Nakamura, K. Endoh, J. Konishi, M.S. Sasaki, T. Miro, S. Fujita, and J.B. Cologne. 1992. Frequency of mutant T lymphocytes defective in the expression of the T-cell antigen receptor gene among radiation-exposed people. Mutation Research 265:173-180.

Laerum, O. D., and R. Bjerknes, eds. 1992. Flow Cytometry in Hematology. Analytical Cytology Series. London, England: Academic Press Limited.

Langlois, R. G., B. A. Nisbet, W. L. Bigbee, D. N. Ridinger, and R. H. Jensen. 1990. An improved flow cytometric assay for somatic mutations at the Glycophorin A locus in humans. Cytometry 11:513-521.

Maier, P., and H. P. Schwalder. 1986. A two-parameter flow cytometry protocol for the detection and characterization of the clastogenic, cytostatic and cytotoxic activities of chemicals. Mutation Research 164:369-379.

Marrone, B. L., D. J. Simpson, T. M. Yoshida, C. J. Unkefer, T. W. Whaley, and T. N. Buican. 1991. Single cell endocrinology: Analysis of P-450scc activity by fluorescence detection methods. Endocrinology 128:2654-2656.

Martin, J. C., and D. E. Swartzendruber. 1980. Time: A new parameter for kinetic measurements in flow cytometry. Science 207:199-200.

Rauchman, M. I., J. C. Wasserman, D. M. Cohen, D. L. Perkins, S. C. Hebert, E. Milford, and S. R. Gullans. 1992. Expression of GLUT-2 cDNA in human B lymphocytes: Analysis of glucose transport using flow cytometry. Biochimica et Biophysica Acta 1111:231-238.

Rhein, A. P., K. P. Gilbertz, and D. van Beuningen. 1994. Analysis of early changes in cell cycle kinetics after low dose irradiation. Presented at the Rocky Mountain Cytometry Conference. 27-29 April 1994, Angel Fire, New Mexico (USA).

Salzman, G. C., S. B. Singham, R. G. Johnston, and C.F. Bohren. 1990. Light Scattering and Cytometry. Pp. 81-108 in: Flow Cytometry and Sorting. M. R. Melamed, T. Lindmo, and M. L. Mendelsohn, eds. New York, NY: Wiley-Liss.

Saunders, G. C., J. H. Jett, and J. C. Martin. 1985. Amplified flow-cytometric separation-free fluorescence immunoassays. Clinical Chemistry 31:2020-2023.

Schreiber, G. A., W. Beisker, M. Bauchinger, and M. Nusse. 1992. Multiparametric flow cytometric analysis of radiation-induced micronuclei in mammalian cell cultures. Cytometry 13:90-102.

Shapiro, H. M. 1988. Practical Flow Cytometry, Second Edition. New York, NY: Wiley-Liss.

Steinkamp, J. A., and H. A. Crissman. 1993. Resolution of fluorescence signals from cells labeled with fluorochromes having different lifetimes by phase-sensitive flow cytometry. Cytometry 14:210-216.

Steinkamp, J. A., T. M. Yoshida, and J. C. Martin. 1993. Flow cytometer for resolving signals from heterogeneous fluorescence emissions and quantifying lifetime in fluorochrome-labeled particles by phase-sensitive detection. Rev. Sci. Instrum. 64:3440-3450.

VanDilla, M. A., L. L. Deaven, K. L. Albright, et al. 1986. Human chromosome-specific DNA libraries: construction and availability. Bio/Technology 4:537-552.

Watson, J. V. 1991. Introduction to Flow Cytometry. Cambridge, England: Cambridge University Press.

Welleweerd, J., M. E. Wilder, S. G. Carpenter, and M. R. Raju. 1984. Flow cytometric determination of radiation induced chromosome damage and its correlation with cell survival. Radiation Research 99:44-51.

Wheeless Jr., L. L. 1990. Slit-Scanning. Pp. 109-126 in: Flow Cytometry and Sorting. M. R. Melamed, T. Lindmo, and M. L. Mendelsohn, eds. New York, NY: Wiley-Liss.

Wilkerson, C. W., P. M. Goodwin, W. P. Ambrose, J. C. Martin, and R. A. Keller. 1993. Detection and lifetime measurements of single molecules in flowing sample streams by laser induced fluorescence. Applied Physics Letter 62:2030-2035.

CASES IN POINT: MONITORING WORKER EXPOSURES TO METALS

Biomarkers and Occupational Health: Progress and Perspectives. 1995. Pp. 293-303

A Genetic Marker for Chronic Beryllium Disease

Cesare Saltini

Exposure to beryllium dusts and fumes has long been recognized as a cause of lung disease (Powers, 1991). While acute exposure to high levels of beryllium (usually exceeding 100 $\mu g/m^3$) has been associated with a form of acute pneumonia (Van Ordstrand et al., 1943), chronic exposure even at significantly lower doses has been associated with a chronic interstitial lung disorder called chronic beryllium disease (CBD) or berylliosis (Tepper et al., 1961). The implementation of a recommendation by the Atomic Energy Commission in 1949 to reduce beryllium levels in the workplace to 2 $\mu g/m^3$ (Sterner and Eisenbud, 1951), has lead to the elimination of acute pneumonia from the workplace in the United States. However, CBD remains the major occupational hazard in beryllium manufacturing (Kriebel et al., 1988).

CBD is characterized by chronic inflammation of the lower respiratory tract. The inflammation is maintained by lymphocyte and mononuclear phagocyte accumulation, granuloma formation, and the deposition of fibrotic tissue. Typical of interstitial lung disorders (Crystal et al., 1984), both the accumulation of mononuclear cells in the interstitium and the deposition of fibrotic tissue contribute to the alteration of the lung anatomical structures which is followed by progressive respiratory dysfunction. Although the clinical course of chronic beryllium disease may be very slow, it can eventually lead to dysfunction and severe disability (Figure 1) requiring treatment with chronic immunosuppressive therapy and, eventually, the need for lung transplantation.

The work of Sterner and Eisenbud (1951) showed that the incidence of chronic beryllium disease is not strictly dose-dependent, thus suggesting that the disease might be due to hypersensitivity to beryllium. Several lines of evidence have since indicated that the granulomatous inflammation of chronic beryllium disease is in fact maintained by an antigen (hapten)-specific reaction of T-cells to beryllium. All individuals with CBD have blood and/or lung T-cells that react against beryllium (Epstein et al., 1982; Pinkston et al., 1984). *In vitro* evaluations have demonstrated that blood and lung T-cells from patients with CBD proliferate in response to the ion, while those from normal individuals do not (Rossman et al., 1988). In this regard, cytofluorimetric analysis of beryllium-induced lung T-cell proliferation suggests that the number of T-cells specifically reacting to beryllium in the lungs of affected individuals may be as high as 10% (Saltini et al., 1989; 1990).

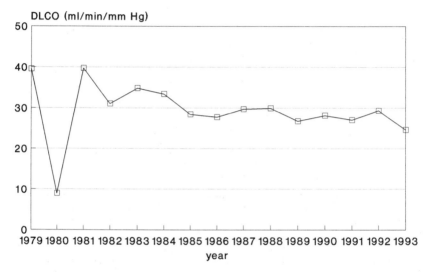

FIGURE 1. Lung function abnormalities of a beryllium-exposed individual with CBD. Shown are changes in lung diffusion capacity.

T-cell clones, which are considered to be representative of individual T-cells involved in the immune response to a specific antigen, were used to clarify the mechanisms of the response of lung T-cells to beryllium. T-cell lines were generated by *in vitro* stimulation with beryllium from T-cells obtained from the lungs of individuals with CBD. These beryllium-induced lung T-cell lines and clones were specific for beryllium in that they did not respond to other metals such as nickel or zirconium, which are known to be able to stimulate T-cells, nor did they respond to the recall antigens, such as tetanus and streptokinase, to which the patients were sensitized (Saltini et al., 1989).

The second important observation made with these T-cell lines was that most of the cells responding to beryllium were CD4$^+$ cells, i.e., helper cells. While the CD8$^+$ lymphocytes suppressor/cytotoxic cells are capable of killing cells infected by viruses or tagged by chemical haptens, the CD4$^+$ helper cells produce interleukin-2, the lymphocyte growth factor, and therefore, induce granuloma formation. In addition, CD4$^+$ cells recognize antigens absorbed or posted on the surface of specialized presenting cells such as the lung macrophages and dendritic cells. The T-cell has a receptor, called the T-cell antigen receptor, which is capable of recognizing peptide antigens, i.e., small fragments of protein bound by a special antigen-presenting cell surface molecules, the HLA molecule. The interaction between the T-cell receptor, the HLA molecule and the bound peptide, with the help of the CD4 molecule, induces the activation of the T-cell and triggers cell proliferation. T-cells can recognize hapten-tagged peptides, i.e., peptides with chemical or metal groups attached to amino acid residues, as well as they recognize natural peptides (Martin et al., 1992).

In vitro experiments using beryllium-specific T-cell lines were designed to investigate the basic mechanism of beryllium recognition. T-cell lines were stimulated with beryllium in the presence of monoclonal blocking antibodies specific either for HLA class I molecules (i.e., those HLA molecules, HLA-A, -B and -C, that present antigens to the CD8$^+$ cytotoxic T-cells) or for class II HLA molecules (i.e., those HLA molecules, HLA-DR, -DQ and -DP, that present antigens to the CD4$^+$ cells). Beryllium-stimulated T-cell line proliferation was blocked only by the anti-MHC class II antibody. This means that, consistent with the CD4$^+$ T-cell phenotype, beryllium recognition by T-cells is mediated by its presentation in the context of HLA-class II molecules (DR or the DQ or DP), but not of HLA-class I molecules, i.e., as a true antigen (Saltini et al., 1989). Further, in a preliminary experiment aimed at identifying the HLA class II molecules responsible for presenting beryllium, T-cell clones stimulated with beryllium in the presence of antigen-presenting cells from individuals expressing the same HLA-DP alleles as the donor of the T-cell line, responded to the metal, while they did not do so when stimulated in the presence of antigen presenting cells expressing the same HLA-DR allele of the donor. This suggests that HLA-DP molecules may be directly involved in presentation of beryllium to T-cells.

These experiments demonstrate that CD4$^+$ T-cell lines from CBD-affected persons recognize beryllium presented by the major histocompatibility complex, or HLA molecules. This observation is similar to those related to other metals that have been recognized as capable of inducing allergy, such as nickel or gold. In the response to nickel, nickel binds to proteins and peptides and the nickel-peptide complex is recognized by T-cells from individuals with nickel allergy. In the reaction to gold, gold binds directly to the HLA molecules and is then recognized by gold-hypersensitive individuals treated with crysotherapy for rheumatoid arthritis (Romagnoli et al., 1992).

As for other metals, the reaction to beryllium is reminiscent of autoimmune reactions in which autologous proteins become recognized as foreign by the immune system. The MHC or HLA molecules, the products of highly polymorphic genes, are responsible for binding and presenting antigens to the T-cells. The affinity of peptide binding varies for different allelic forms of these molecules which can therefore select for the antigen repertoire recognized by each individual. Susceptibility to immune reactions is associated with the allelic variants of the HLA molecules. HLA-DQ alleles expressing an amino acid different from the aspartic acid in position 57 of the $\beta 1$ chain are associated with susceptibility to diabetes, HLA-DR alleles with rheumatoid arthritis, and HLA-DP alleles with juvenile rheumatoid arthritis, suggesting that these allelic variants may play a crucial role in the reaction against the self-antigens which become recognized by T-cells as foreign targets (Bjorkman et al., 1987; Todd et al., 1988).

RESULTS

With this background, we asked whether susceptibility to beryllium disease may be associated with HLA allelic variants, namely with HLA-DP, which could be the best candidate for presentation of beryllium to T-cells. In the MHC locus, where all the HLA genes are located, the HLA-DP gene is located at a distant site from the DQ and the DR genes and it may be inherited independently of them. In this context, while diseases associated with HLA-DQ and DR may in fact be associated with both of them jointly, a disease associated with HLA-DP is likely to be directly and uniquely associated with this gene.

The structure of the HLA genes is characterized by the high number of polymorphisms and the similarly high number of allelic variants. Mutations generating polymorphisms that are important for antigen presentation have accumulated in defined regions of the HLA genes.

HLA gene allelic variants differ at definite hypervariable regions of the DNA sequence. The 32 alleles of the HLA-DP gene identified so far differ at 6 hypervariable regions of the exon 2 sequence (Marsh and Bodmer, 1992).

To evaluate whether susceptibility to CBD is associated with HLA-DP, a population of beryllium disease patients was examined for biases in the frequency of individual HLA-DP alleles using several complementary molecular typing techniques. A sample of exposed individuals with CBD, identified retrospectively by clinical symptoms, X-ray abnormalities, pulmonary function abnormalities, a positive lung tissue biopsy showing granulomas, and a positive *in vitro* lymphocyte proliferation test response to beryllium (BeLPT), were evaluated in comparison to a sample of individuals exposed to beryllium but having a negative BeLPT.

HLA-DP alleles were identified by heteroduplex analysis and confirmed using allele specific oligonucleotides, i.e., probes specific for each of the six HLA-DP sequence variable regions. This analysis showed that there was a marked prevalence of the HLA-DP 2 allele in the patient population, compared to the exposed population, along with a comparable prevalence of the HLA-DP 4 allele in the unaffected, exposed population. The HLA-DP allelic frequencies of this population were not different from the frequencies of normal populations obtained by two independent studies (Richeldi et al., 1993). Having identified two alleles which were positively and negatively associated with the disease, the frequencies of sequence polymorphisms for which these two alleles differ were examined in the affected and unaffected populations using allele-specific DNA amplification with an amplification refractory mutation system, and direct PCR sequencing of the 200 bases of the second exon of the HLA-DPβ1 gene. Looking at polymorphisms in region C, where the HLA-DP 2 and the DP 4 alleles differ for two amino acid-coding changes, and in region D, where they differ for one amino acid-coding change, we found that the glutamic acid position 69-coding polymorphism in region D was expressed in 32 of 33 patients (97% of the patient sample). Strikingly, only 27% of the control population had this polymorphism. In addition to HLA-DP 2 (DPB1*0201 and *0202) alleles, which represented most of the glutamic 69-positive alleles, there were rarer alleles such as the HLA-DPB1*0901, *1001, *1301 and *1901 which were over-represented in this patient population sample (Richeldi et al., 1993).

In autoimmune disorders, which show some similarities with metal allergies, diseases have been found to be associated with normal HLA alleles, and not with new mutations induced by autoimmunity inducing agents. Patients with rheumatoid arthritis, for example, do not express abnormally mutated alleles; they have normal HLA-DR 4 molecules which are probably more prone to bind certain self-molecules and present them to T-cells in the synovium. The same is thought to happen with diabetes (Bjorkman et al., 1987; Todd et al., 1988). The only known exception is that of the non-obese diabetic (NOD) mouse which has a mutated IA (I-A^{g7}) molecule that is homologous with the human HLA-DQB1*0302 molecule (Todd et al., 1991). There have been thousands of chromosomes evaluated in autoimmune and inflammatory disorder association studies, and the HLA sequences obtained seem to belong to normal alleles. Looking at the population sample with beryllium disease, HLA-DPβ1 exon II sequences obtained with direct PCR sequencing were normal allelic sequences, including the glutamate 69-coding polymorphism in 27 out of 28 sequences examined. Thus, the CBD-associated HLA-DP sequence is not a new mutation arising from beryllium exposure, but a normal HLA-DP allelic variant, which may be endowed with a greater ability to react to certain stimuli, possibly to beryllium (Richeldi et al., 1993).

In summary, beryllium-specific CD4$^+$ cells recognized beryllium in association with HLA class II molecules at the surface of the antigen presenting cells. This response may be restricted by the HLA-DP molecule itself. CBD is strongly associated with a single amino acid variant (glutamic acid-69) in the β1 chain of the HLA-DP molecule. No unknown mutated variants have been recognized in the patient sample evaluated so far.

DISCUSSION

Are there other susceptibility genes in addition to HLA-DP, as is the case in autoimmune disorders such as autoimmune diabetes? The HLA-DRβ3 (HLA-DRw53) and TNF genes have been examined so far, and no linkage with CBD has been found (Richeldi et al., 1993). This may suggest that, as in autoimmune diabetes, the HLA gene (HLA-DP in this case) may be the primary susceptibility gene for disease. It is likely that other susceptibility complementing genes do exist in beryllium disease as well, although their identification is not likely to be a simple task.

What are the implications of these findings for CBD management and prevention? We have found that an HLA marker is associated with a disease caused by the reaction to beryllium. Based upon these data, one may imagine that if the macrophage in the lung of an exposed individual with the "right" HLA-DP alleles is tagged with beryllium, it can present beryllium to lung T-cells that recognize a beryllium-modified HLA-DP molecule or a beryllium-modified HLA-DP-bound peptide. This may be sufficient to induce an abnormal immune reaction and disease. However, the incidence of disease has always been found to be less than 5%, while the frequency of this HLA-DP marker in the general population is about 27%. Hence, there are probably other factors and some other interactions besides this hypothetical interaction between the HLA-DP molecule and beryllium-recognizing T-cells. It may be cytokine, cytokine receptor genes or cell proliferation-modulating oncogenes which may favor inflammatory processes that are involved in increasing the susceptibility to beryllium in that 5% of exposed individuals who do eventually develop the disease. Whatever the interaction between the environmental factors and the genetic factors that may be involved in susceptibility to beryllium disease, what is the potential role of this genetic marker, which may have a 10-20% predictive power, in the diagnosis and treatment of CBD?

We know from the studies on BeLPT testing done by Rossman et al. (1988), Newman et al. (1989), and Kreiss et al. (1993) that while there are patients who clearly have active disease as diagnosed by exposure, positive granulomas, and abnormal X-rays and by pulmonary function (i.e., the type of patient identified for our studies on HLA-DP association), there are also patients who are being identified by a positive BeLPT who have positive granulomas in their lungs but do not have any clinical, chest X-ray or pulmonary function abnormalities. Further, there are individuals identified by a positive BeLPT, who do not even have detectable granulomas in their biopsies (Kreiss et al., 1994). In the context of the studies with the HLA-DP genetic marker, this observation brings up some questions. We do not know yet whether these individuals, who may have been diagnosed at an earlier stage of disease or may have a different and more benign form of beryllium allergy, have the same genetic background as those diagnosed with clinically active disease. Some rather large studies are necessary to evaluate the genetics of this allergy-positive disease-negative population, in relation to the HLA-DP glutamic 69 marker. Should they have the same genetic background, it would be rational to use genetic markers to identify exposed individuals who need stricter immunologic follow-up.

On the other hand, immunology is rapidly moving toward the use of targeted treatment of autoimmune disorders using the immunogenetic data to identify either blocking antibodies directed at the specific site of the autoimmune reaction or at the competing autoimmune peptide antigen. In this regard, a more detailed understanding of the precise mechanism of the interaction between the HLA-DP molecule, the T-cell and beryllium will be instrumental to the design of therapeutic strategies (Figure 2).

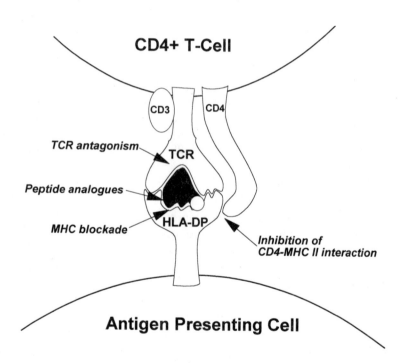

FIGURE 2. Schematic representation of the interaction between the T-cell receptor (TCR) of a CD4$^+$ T-cell and the MHC class II molecule (in this case HLA-DP) of an antigen presenting cell. Also shown is the peptide antigen (in black). Based upon the mechanism of antigen presentation, some possible therapeutic strategies are indicated (see text).

Chronic beryllium disease is thus a disease model for which a genetic disease susceptibility marker can help not only with understanding the pathogenesis of disease but also with early diagnosis and prevention.

This is particularly important since recent observations of other occupational allergies such as acid anhydride-induced asthma (Young et al., 1993) and isocyanate-induced asthma (Bignon et al., 1994) may also suggest roles for the HLA-DR and HLA-DQ genes as well in susceptibility to environmental and occupational factors. With the most recent advances in immunogenetic therapy for autoimmune disorders, these markers of susceptibility to occupational disorders might eventually be used to design newer and more efficacious treatments of environmental allergies.

REFERENCES

Bjorkman, P. J., M. A. Saper, B. Samraoui, W. S. Bennett, J. L. Strominger, and D.C. Wiley. 1987. Structure of the human class I histocompatibility antigen, HLA-A2. Nature 329:506-512.

Bignon, J. S., Y. Aron, L. Y. Ju, M. C. Kopferschmitt, R. Garnier, C. Mapp, L. M. Fabbri, G. Pauli, A. Lockhart, D. Charron, and E. Swierczewski. 1994. HLA class II alleles in isocyanate-induced asthma. American Journal of Respiratory and Critical Care Medicine 149:71-75.

Crystal, R. G., P. B. Bitterman, S. I. Rennard, A. Hance, and B. A. Keogh. 1984. Interstitial lung disease of unknown cause: disorders characterized by chronic inflammation of the lower respiratory tract. New England Journal of Medicine 310:235-244.

Epstein, P. E., J. H. Dauber, M. D. Rossman, and R. P. Daniele. 1982. Bronchoalveolar lavage in a patient with chronic berylliosis: evidence for hypersensitivity pneumonitis. Annals of Internal Medicine 97:213-216.

Kreiss, K., S. Wasserman, M. M. Mroz, and L. S. Newman. 1993. Beryllium disease screening in the ceramics industry. Journal of Occupational Medicine 35:267-274.

Kreiss, K., F. Miller, L. S. Newman, E. A. Ojo-Amaize, M. Rossman, and C. Saltini. 1994. Chronic beryllium disease — from the work place to cellular immunology, molecular immunogenetics, and back. Cell Immunology (submitted).

Kriebel, D., J. D. Brain, N. L. Sprince, and H. Kazemi. 1988. The pulmonary toxicity of beryllium. American Review of Respiratory Diseases 137:464-73.

Marsh, S. G. E. and J. G. Bodmer. 1992. HLA Class II nucleotide sequences. Tissue Antigens 40:229-243.

Martin, S., B. Ortmann, U. Pfllugfelder, U. Birsner, and H. U. Weltzien. 1992. The role of TNP-anchoring peptides in defining hapten-epitopes for MHC-restricted cytotoxic T cells: crossreactive TNP-determinants on different peptides. Journal of Immunology 149:2569-2575.

Newman, L. S., K. Kreiss, T. E. King, S. Seay, and P. A. Campbell. 1989. Pathologic and immunologic alterations in early stages of beryllium disease. American Review of Respiratory Diseases 139:1479-1486.

Pinkston, P., P. Bitterman, and R. G. Crystal. 1984. Interleukin-2 in the alveolitis of beryllium induced lung disease. American Review of Respiratory Diseases 129:A161.

Powers, M. B. History of beryllium. 1991. Pp. 9-24 in: Beryllium. Biomedical and environmental aspects. M. D. Rossman, O. P. Preuss and M. B. Powers, eds. Williams & Wilkins, Baltimore.

Richeldi, L., R. Sorentino, and C. Saltini. 1993. HLA-DP 1 Glutamate 69: a genetic marker of beryllium disease. Science 262:242-244.

Romagnoli, P., G. A. Spinas, and F. Sinigaglia. 1992. Gold-specific T Cells in rheumatoid arthritis patients treated with gold. Journal of Clinical Investigation 89:254-258.

Rossman, M. D., J. A. Kern, J. A. Elias, M. R. Cullen, P. E. Epstein, O. P. Preuss, T. N. Markham, and R. P. Daniele. 1988. Proliferative response of bronchoalveolar lymphocytes to beryllium. Annals of Internal Medicine 108:687-693.

Saltini, C., K. Winestock, M. Kirby, P. Pinkston, and R. G. Crystal. 1989. Maintenance of alveolitis in patients with chronic beryllium disease by beryllium-specific helper T cells. New England Journal of Medicine 320:1103-1109.

Saltini, C., M. Kirby, B. Trapnell, N. Tamura, and R. G. Crystal. 1990. Biased accumulation of T-lymphocytes with "memory"-type CD45 leukocyte common antigen gene expression on the epithelial surface of the human hung. Journal of Experimental Medicine 171:1123-1140.

Sterner, J., and M. Eisenbud. 1951. Epidemiology of beryllium intoxication. Archives of Industrial Hygiene and Occupational Medicine 4:123-157.

Tepper, L. B., H. L. Hardy, and R. I. Chamberlin. 1961. Toxicity of beryllium compounds. Elsevier Publishing Company, Amsterdam.

Todd, J. A., H. Acha-Orbea, and J. I. Bell. 1988. A molecular basis for MHC class II-associated autoimmunity. Science 240:1003-1009.

Todd, J. A., T. J. Aitman, R. J. Cornall, S. Ghosh, J. R. S. Hall, C. M. Hearne, A. M. Knight, J. M. Love, M. A. McAleer, J. B. Prins, N. Rodrigues, M. Lathrop, A. Pressey, N. H. DeLarato, L. B. Peterson, and L. S. Wicker. 1991. Genetic analysis of autoimmune type 1 diabetes mellitus in mice. Nature 351:542-547.

Van Ordstrand, H. S., R. Hughes, and M. G. Carmody. 1943. Chemical pneumonia in workers extracting beryllium oxide. Cleveland Clinical Quarterly 10:10-18.

Young, R. P., R. Barker, W. Cookson, and A. J. Newman Taylor. 1993. HLA-DR and DP antigen frequencies and the anhydride sensitization. American Review of Respiratory Disease 147:A113.

Biomarkers and Occupational Health: Progress and Perspectives. 1995. Pp. 304-312

Immunology of Chronic Beryllium Disease

Milton D. Rossman

Chronic beryllium disease was first described by Hardy and Tabershaw (1946) at the Massachusetts General Hospital. In this publication, the disease was referred to as delayed chemical pneumonitis. The first 17 cases were contrasted to acute beryllium disease which was thought then (and still is thought) to be an acute chemical pneumonitis (Van Ordstrand et al., 1943). The symptoms in these cases began at least six months after initial exposure and included anorexia, weight loss, cough, and dyspnea on exertion. On physical examination, fever, tachycardia and rales were common, whereas cyanosis, edema, and clubbing (evidence of extensive pulmonary disease) were present in less than 30 percent of the cases. While chronic beryllium disease is predominately a pulmonary granulomatous disease, granulomas have also been described in the skin, liver, spleen, hilar and mediastinal lymph nodes, and the conjunctiva. Hypercalcemia and nephrocalcinosis have also been reported (Stoeckle et al., 1969). In contrast to sarcoidosis, posterior uveitis, central nervous system effects and myocardial disease are thought not to occur in chronic beryllium disease.

During industrial processing, beryllium can be aerosolized either as a vapor mist or dust. After inhalation and deposition in the lung, beryllium will migrate to the draining lymph nodes and into the blood stream. The circulating beryllium will either be excreted in the urine or deposited into other tissues. The half-life for clearance of beryllium from the lungs has been measured in animals (Reeves, 1991). While a portion of the beryllium is cleared rapidly, some beryllium is cleared very slowly with a half-life that is measured in months to years.

Two aspects of chronic beryllium disease are responsible for the slow progress that has been made in understanding the pathogenesis of this disease. First, the incidence of chronic beryllium disease, even in heavily exposed populations, is low. Before adequate environmental controls were in place, the incidence of chronic beryllium disease in exposed foundry, machining, and extraction workers ranged from less than 1 percent to 4.9 percent (Eisenbud and Lisson, 1983). The second aspect of this condition that has slowed progress is the long latency period between an individual's first exposure and the clinical recognition of disease. The mean latency period is 10 years, but some cases have not been recognized until 40 years after the initial exposure.

Radiographic abnormalities of chronic beryllium disease are non-specific and cannot be differentiated from sarcoidosis (Sargent, 1991; Aronchick et al., 1987). Small rounded nodules are more frequently found in the upper lobes but may involve all lung zones. Calcification of the nodules can occur, and the nodules may develop into large opacities that resemble progressive massive fibrosis. In the advanced stages of the disease, fibrosis and contraction of the upper lobes as well as scar emphysema may occur. Hilar adenopathy may occur in up to 50 percent of cases and frequently will be mild to moderate in extent. Dramatic clearing in the radiographic evidence of the disease has been observed with corticosteroid therapy (Aronchick et al., 1987). Biopsy-proven cases of pulmonary chronic beryllium disease have been described with normal chest radiographs. Computed tomography has confirmed the standard chest radiographic description of chronic beryllium disease and documented the non-specific nature of the abnormalities (Harris et al., 1993). It is not known whether standard or high resolution computed tomography will be able to detect all cases of chronic beryllium disease.

The classic histology of chronic beryllium disease is the noncaseating granuloma (Jones-Williams, 1991). Epithelioid cells may contain Schaumann bodies (calcified inclusions that do not contain beryllium) and/or asteroid bodies. Multi-nucleated giant cells are frequent, and there is a surrounding cuff of lymphocytes. This inflammatory process should be completely reversible with corticosteroid therapy. However, in untreated progressive disease, fibroblasts invade the granuloma with deposition of reticulin and collagen. Denatured collagen will persist as hyalinized granuloma. Once this scarring and destruction have occurred, the changes are irreversible.

Immunology

The suspicion that chronic beryllium disease was due to an immunologic reaction to beryllium was based on the following observations: 1) Beryllium painted on the skin (patch testing) could elicit delayed-type hypersensitivity reactions in patients with chronic beryllium disease or sensitize some individuals who had never been exposed (Curtis, 1951). However, because of the concern that patch testing could sensitize individuals to beryllium, skin testing has not been widely used; 2) Only 5 percent of heavily exposed workers have developed the disease despite large differences in exposure histories (Eisenbud and Lisson, 1983); 3) Chronic beryllium disease was associated with "immunologic granuloma"; and, 4) Animal studies demonstrated that a hypersensitivity could be elicited in animals (Alekseeva, 1965).

In the 1970s, *in vitro* studies that simulated patch testing were developed and applied to patients with chronic beryllium disease. Although these studies were originally called lymphocyte transformation tests, they were actually measuring the normal growth and proliferation of lymphocytes after antigen exposure. Today these assays are referred to as lymphocyte proliferation tests. The blood cells from a large percentage of patients with chronic beryllium disease had positive proliferative responses to beryllium (Marx and Burrell, 1973; Deodhar et al., 1973). In addition, activated lymphocytes will secrete proteins/peptides (lymphokines) that activate macrophages and other inflammatory cells. After stimulation with beryllium, blood cells from many patients with chronic beryllium disease released a lymphokine, such as macrophage inhibition factor (Williams and Jones-Williams, 1982). However, the significance of these results was uncertain because all patients with chronic beryllium disease did not have positive responses. Thus, whether chronic beryllium disease was due to an abnormal immunologic reaction, and/or whether the immunologic reaction was an epi-phenomenon were still in question.

The confirmation that chronic beryllium disease was due to a cell-mediated response to beryllium came in the 1980s when cells harvested from the bronchoalveolar lavage fluid from patients with chronic beryllium disease were examined (Rossman et al., 1988; Epstein et al., 1982; Cullen et al., 1987). A marked increase in the number and percent of lymphocytes in the bronchoalveolar lavage fluid was observed [normal = 10.9 ± 2.0 (mean percent $\pm$ SE); CBD = 56.8 ± 6.7; $p < 0.05$]. When analyzed by flow cytometry, these cells were noted to be predominantly T lymphocytes ($80.3\% \pm 5.8$) and of the CD4+ phenotype ($67.8\% \pm 4.9$). Thus, patients with chronic beryllium disease had a CD4+ T cell lymphocytic alveolitis similar to sarcoidosis, a finding consistent with a hypersensitivity reaction.

However, major evidence that a chronic cell-mediated response to beryllium was the major pathogenic mechanism for chronic beryllium

disease was observed when beryllium proliferation studies were conducted with cells collected from the bronchoalveolar lavage fluid of patients with chronic beryllium disease (Rossman et al., 1988). The cells
from 14 consecutive patients with pathologically confirmed granulomas
and definite exposure to beryllium had positive responses. None of
the patients with negative proliferative responses of their lung cells to
beryllium had chronic beryllium disease. This included 16 patients with
sarcoidosis and no history of beryllium exposure and 4 beryllium workers with biopsy evidence of non-beryllium lung disease. Not only did all
patients with chronic beryllium disease have a positive proliferative response of their bronchoalveolar cells to beryllium, but this response was
more pronounced in their lung cells than their blood cells. Thus, there
was an accumulation of beryllium specific cells in the lungs of patients
with chronic beryllium disease (Table 1). The proliferative responses of
these cells exhibited a dose-response relationship to beryllium with the
peak response to 10 to 100 μM beryllium sulfate or beryllium fluoride.
In addition, the peak response occurred after 3 to 5 days of *in vitro* cell
culture.

TABLE 1. COMPARISON OF PEAK BLOOD AND LUNG PROLIFERATIVE
RESPONSES IN A PATIENT WITH CHRONIC BERYLLIUM DISEASE

Stimulus	Blood*	Lung*
PHA	169	41
Con A	57	20
Candida	14	5
Tetanus	3	6
BeSO4	2	60
BeF2	2	55

* Values expressed as stimulation index (Test CPM/Control CPM)

To further understand the mechanism of beryllium stimulation of
blood and lung lymphocytes, Saltini et al. (1989) observed that only
CD4+ T cells responded to beryllium *in vitro*. The beryllium response
could be blocked by antibodies against the IL-2 receptor on T cells or
against MHC Class II molecules. This suggested that the beryllium-
induced lymphocyte proliferative response was a normal immunological

response since the T cell receptor on CD4+ T cells recognizes antigen
in the context of MHC Class II molecules. The beryllium-sensitive T
cells could be cloned and retained specific reactivity for beryllium but
not for other antigens. In addition, when cells were cloned from the
same patients to other antigens, these cells did not react to beryllium.
By Southern analysis, these beryllium-specific T cell clones contained
different T cell antigen receptors. This implied that the T cell response
to beryllium was not a clonal proliferation but a polyclonal response.

T Cell Receptors

In collaboration with Drs. Williams and Weiner at the University
of Pennsylvania, we began to define the specific T cell proteins that
compose the T cell antigen receptor. The T cell antigen receptor consists
of an acidic or α chain and a basic or β chain. Because the T cell antigen
receptor must have the potential to recognize all of the foreign antigens,
a complex method of somatic recombination occurs so that a limited
number of gene segments can recombine in different patterns that can
result in more than a million different protein products. The process of
somatic recombination occurs in the thymus as the T cell matures. For
the α chain, there are multiple gene segments that code for the variable
(V) and the joining (J) region but only a single constant (C) segment.
For the β chain, in addition to the V and J gene segments, there are also
diversity (D) segments and 2 constant (C) gene segments. In addition,
nucleotide additions can be added between the V, D, and J segments.
For simplicity, the α and β chains can each be grouped into 20 or more
subfamilies based on V gene usage. When the T cell emerges from the
thymus, each T cell expresses a single form of the T cell receptor on
its surface. Thus, a different T cell clone exists for each T cell receptor
expressed.

From 8 patients with chronic beryllium disease, the RNA was ex-
tracted and converted to cDNA. Utilizing DNA probes specific for $18V\alpha$
subfamilies and $19V\beta$ subfamilies, we were consistently able to amplify
the cDNA using PCR from the $V\alpha2$ and the $V\beta7$ and $V\beta12$ subfamilies
from the freshly isolated lung cells of patients with chronic beryllium
disease. To determine whether the increased expression of these specific subfamilies was polyclonal or oligoclonal, we cloned the cDNA into
a vector and then sequenced the T cell receptor α or β chains. In
each case, an oligoclonal distribution of T cell antigen receptor chains
was observed. The oligoclonal distribution of T cell receptor chains is

consistent with a typical antigen-induced response rather then a super-antigenic type response.

Thus, analysis of the T cell antigen receptor sequences is also consistent with beryllium acting as a typical antigen. It is presumed that beryllium must be acting as a hapten and binds to an as yet unidentified peptide. This beryllium-peptide complex will be presented by MHC Class II molecules to the T cell antigen receptor.

MHC Class II Molecules

Since there are 3 types of Class II molecules, DR, DQ and DP, in collaboration with Dr. Elizabeth Mellins, we attempted to determine which Class II molecules were presenting the beryllium-peptide complex to the T cell. Using monoclonal antibodies against DR, L243 and DP, B7/21, the MHC restriction of freshly isolated peripheral blood cells was determined. In 3 consecutive patients, complete inhibition of the beryllium-induced proliferative response was observed with the anti-DR antibody. In contrast, almost no inhibition was noted with the anti-DP antibody. These studies suggest that DR molecules are important in presenting the beryllium-peptide complex to the T cell antigen receptor while DP molecules have little or no role in presentation.

These results are especially interesting because of the recent studies by Richeldi et al. (1993) that analyzed MHC Class II molecules in patients with chronic beryllium disease. Preliminary analysis revealed no association with DR or DQ. However, using heteroduplex analysis, certain DP alleles were associated with chronic beryllium disease. The HLA-DPB1*0201 allele was positively associated with beryllium disease while the HLA-DPB1*0401 allele was negatively associated with beryllium disease. When glutamic acid is found at position 69, which is the case for the HLA-DPB1*0201 allele but not for the HLA-DPB1*0401 allele, 97% of 33 patients with chronic beryllium disease expressed this pattern while only 30% of the control population expressed the same motif.

In order to confirm this observation, in collaboration with Dr. Dimitri Monos, we also performed molecular typing of DP molecules using sequence specific oligonucleotide probes (SSOP). Twenty-nine patients with evidence of beryllium hypersensitivity, i.e., evidence of blood or bronchoalveolar cells with positive proliferative responses to beryllium, were compared to 84 individuals with beryllium exposure but no evidence of hypersensitivity. In contrast to the previous findings, no association with HLA-DPB1*0201 was observed, but a positive association with HLA-DPB1*0601 was observed. Similar to the previous study, we also observed that HLA-DPB1*0401 was protective. Interestingly, glutamic acid at position 69 is also present on HLA-DPB1*0601. When the frequency of this motif was determined, glutamic acid at position 69 was observed in 24 of 29 cases. Thus, our studies also demonstrated the association of glutamic acid at position 69 with beryllium disease.

The significance of this observation is uncertain. Is this motif a marker for some other closely related allele? The antibody studies would suggest that DP is not involved directly in presenting the beryllium-peptide complex to the T cell receptor, and yet, the association of a specific motif on the DPB1 allele has been associated with beryllium disease in two independent studies. Could this pattern play some other role in the proliferative responses to beryllium?

CONCLUSIONS

Understanding the immunological mechanism involved in beryllium hypersensitivity has greatly aided our understanding of the pathophysiology of chronic beryllium disease. Immunologic testing is already used in screening assays for the early identification of cases (Kreiss et al., 1993). For beryllium disease, this is very important since treatment is available that can arrest and reverse the inflammatory reaction. In addition, the identification of the specific molecules that bind beryllium-peptide complexes on antigen presenting cells and/or the specific T cell antigen receptors on CD4+ T lymphocytes may lead not only to the development of new tests to identify patients with beryllium disease and workers at risk for the development of chronic beryllium disease but also may lead to the development of new therapeutic approaches to this disease.

REFERENCES

Alekseeva, O. G. 1965. [Ability of beryllium compounds to produce delayed allergy.] Gig Trud Prof Zabol 11:20-25.

Aronchick, J. M., M. D. Rossman, and W. T. Miller. 1987. Chronic beryllium disease: diagnosis, radiographic findings, and correlation with pulmonary function tests. Radiology 163:677-682.

Cullen, M. R., J. R. Kominsky, M. D. Rossman, et al. 1987. Chronic beryllium disease in a precious metal refinery: clinical, epidemiologic and immunologic evidence for continuing risk from exposure to low level beryllium fume. American Review of Respiratory Disease 135:201-209.

Curtis, G. H. 1951. Cutaneous hypersensitivity due to beryllium: a study of 13 cases. American Medical Association Archieves of Dermatological Syphilis 64:470-482.

Deodhar, S. D., B. Barna, and H. S. Van Ordstrand. 1973. A study of the immunologic aspects of chronic berylliosis. Chest 63:309-313.

Eisenbud, M., and J. Lisson. 1983. Epidemiological aspects of beryllium - induced nonmalignant lung disease: a 30-year update. Journal of Occupational Medicine 25:196-202.

Epstein, P. E., J. H. Dauber, M. D. Rossman, and R. P. Daniele. 1982. Bronchoalveolar lavage in a patient with chronic berylliosis: evidence for hypersensitivity pneumonitis. Annals of Internal Medicine 97:213-216.

Hardy, H. L., and I. R. Tabershaw. 1946. Delayed chemical pneumonitis occurring in workers exposed to beryllium compounds. Journal of Industrial Hygiene and Toxicology 28:197-211.

Harris, K. M., K. McConnochie, and H. Adams. 1993. The computed tomographic appearances in chronic berylliosis. Clinical Radiology 47:26-31.

Jones-Williams, W. 1991. Pathology and histopathology of chronic beryllium disease. Pp. 151-160. in Beryllium: Biomedical and Environmental Aspects. M. D. Rossman, O. Preuss, and M. Powers, eds. Williams and Wilkins, Baltimore.

Kreiss, K., S. Wasserman, M. M. Mroz, and L. S. Newman. 1993. Beryllium disease screening in the ceramics industry: blood lymphocyte test performance and exposure-disease relations. Journal of Occupational Medicine 35:267-274.

Marx, J. J., Jr., and R. Burrell. 1973. Delayed hypersensitivity to beryllium compounds. Journal of Immunology 111:590-598.

Reeves, A. L. 1991. Toxicokinetics. Pp. 77-86 in Beryllium: Biomedical and Environmental Aspects. M.D. Rossman, O. Preuss, and M. Powers, eds. Williams and Wilkins, Baltimore.

Richeldi, L., R. Sorrentino, and C. Saltini. 1993. HLA-DPB1 Glutamate 69: a genetic marker of beryllium disease. Science 262:242-244.

Rossman, M. D., J. A. Kern, J. A. Elias, M. R. Cullen, P. E. Epstein, O. P. Preuss, T. N. Markham, and R. P. Daniele. 1988. Proliferative response of bronchoalveolar lymphocytes to beryllium. Annals of Internal Medicine 108:687-693.

Saltini, C., D. Winestock, M. Kirby, P. Pinkston, and R. G. Crystal. 1989. Maintenance of alveolitis in patients with chronic beryllium disease by beryllium -specific T cells. New England Journal of Medicine 320:1103-1109.

Sargent, E. N. 1991. Radiological aspects of chronic beryllium disease. Pp. 141-150: in Beryllium: Biomedical and Environmental Aspects. M. D. Rossman, O. Preuss, and M. Powers, eds. Williams and Wilkins, Baltimore.

Stoeckle, J. D., H. L. Hardy, and A. L. Weber. 1969. Chronic beryllium disease: long term follow-up of sixty cases and selective review of the literature. American Journal Medicine 46:545-561.

Van Ordstrand, H. S., R. Hughes, and M. G. Carmody. 1943. Chemical pneumonia in workers extracting beryllium oxide. Report of three cases. Cleveland Clinical Quarterly 10:10-18.

Williams, W. R., and W. Jones-Williams. 1982. A comparison of lymphocyte transformation (BeLT) and macrophage migration inhibition (BeMIF) tests in the detection of beryllium hypersensitivity. Journal of Clinical Pathology 35:684-687.

Biomarkers and Occupational Health: Progress and Perspectives. 1995. Pp. 313-323

Biological Monitoring of Exposure to Hexavalent Chromium in Isolated Erythrocytes

Leopold W. Miksche and J. Lewalter

Biological monitoring can be helpful in controlling individual exposure to chemicals. However, if the wrong parameter is measured, the information may result in a misperception of safety in situations where, in reality, a hazard is present.

For chromium, it is known that exposure to Cr(VI) can lead to health impairment or even to cancer. On the other hand, Cr(III) is an essential element in nutrition. Many studies have, in fact, shown that the predominant form of chromium recovered in blood, tissues, and from cells cultured with Cr(VI) compounds is Cr(III) (Cohen et al., 1993). Within cells, Cr(VI) is reduced to Cr(III), and the intermediate formation of unstable Cr(IV) and Cr(V) ions, and free radical intermediates is considered to be a possible cause of observed DNA damage. The Cr(III) produced by intracellular reduction is bound in the form of insoluble complexes and/or to protein structures.

Although the intracellular reduction of Cr(VI) is required for DNA damage, extracellular reduction decreases chromium entry into the cells and thereby serves as a detoxification mechanism. Using enzymatic and chemical reduction-oxidation reactions, several agents that exist both within the cells and as components of blood plasma may act as reductants. The most important among these are ascorbic acid and glutathione. The majority of Cr(VI) that enters the body via inhalation or ingestion is quickly reduced to Cr(III) and is thus detoxified. Cr(III) will then be excreted mostly in the urine. Considering this detoxification mechanism, it does not seem reasonable to use Cr levels in urine as the only means of biological monitoring in the framework of occupational health surveillance.

The starting point of our studies on Cr(VI) and Cr(III) compounds was the intention to establish a method for measuring the chromium body burden in occupationally exposed persons under workplace condi-

313

tions rather than measuring chromium excretion in urine as a means of estimating exposure to this metal.

In Vitro Studies

In a study by Korallus et al. (1984), the reducing capacity of various individual samples of plasma or serum of non-exposed volunteers was determined. Human plasma has a spontaneous reduction capacity which is able to detoxify Cr(VI) compounds up to a concentration of approximately 2 ppm into Cr(III).

The ability of ascorbic acid to reduce Cr(VI) in blood or serum was quantitatively determined. In a first set of experiments, pooled serum samples were spiked with ascorbic acid to obtain concentrations of 50, 200, 500, and 1000 ppm. Cr(VI) solution was then added. The mixtures were then measured polarographically, and the actual Cr(VI) concentrations were plotted versus the time elapsed. The resulting curves reflect the reduction rates of Cr(VI) in serum at different levels of ascorbic acid as shown in Figure 1.

As expected, these curves indicate a clear increase in reduction velocity with increasing concentrations of ascorbic acid. The curve of the spontaneous reduction of Cr(VI) is presented for comparison.

In a continuation of this study, the behavior of chromium ions was tested with respect to the permeation of the erythrocyte membrane. Results were already available from studies aimed at the determination of erythrocytes using radiolabelled chromate (^{51}Cr). According to these studies, the number of erythrocytes and their correlated volume, which express a dynamic equilibrium between formation and disintegration of erythrocytes, are subject to only minor variation over time. Their mean life span is normally about 115 days. The uptake of Cr(VI) compounds into erythrocytes *in vitro* depends on various factors. The rate of uptake increases with rising temperature. After about 15 minutes under optimum conditions, more than 90% of unchanged Cr(VI) penetrates into the cells both at 22° C and at 37° C.

Studies of erythrocytes labelled with ^{51}Cr(VI) had also shown that radioactivity diminishes by 4 to 8% during the first 24 hours. This "early loss" is attributed to a rapid elimination of a corresponding portion of Cr (intracellularly formed soluble Cr[III] complexes that can penetrate the cell membrane). After that, there is a roughly exponential loss of activity with a half-life of 25 to 40 days (Cohen et al., 1993).

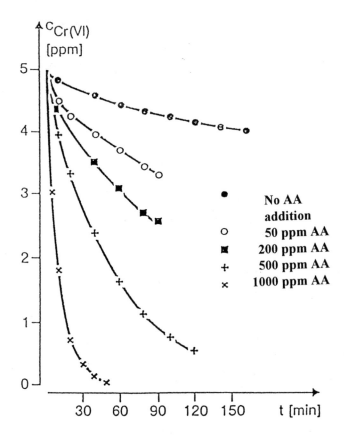

FIGURE 1. Reaction rate of Cr(VI) reduction at different concentrations of ascorbic acid (AA).

Based on these findings, we assumed that Cr determination in erythrocytes might be useful for estimation of the body burden of Cr(VI) following exposure to chromium compounds. This was based on the assumption that only those amounts of Cr(VI) are able to permeate the erythrocyte membrane either because they exceed the plasma reduction capacity (PRC) or because they have entered the erythrocytes before the reduction has occurred. Three questions had to be answered:

- Which concentration of Cr(VI) compounds in whole blood leads to binding of Cr in the erythrocytes?

- What is the effect of additional spiking with ascorbic acid on the permeation of the erythrocyte membrane by Cr(VI) and its binding in the erythrocytes?

- How do Cr(III) compounds normally behave with respect to the permeation of the erythrocyte membrane? Are they also somehow bound in the erythrocytes?

To answer these questions, whole blood samples were spiked with various quantities of highly water-soluble Cr(VI) or Cr(III) compounds, both before and after the addition of ascorbic acid, and the Cr levels in plasma and in the erythrocytes were determined quantitatively (Lewalter et al., 1985).

As shown in Table 1, the following results could be reproduced and confirmed:

- Whole blood samples (specimens 4 and 6) spiked with Cr(VI) show high Cr values not only in plasma, but also in corresponding quantities of erythrocytes.

- Spiking with Cr(VI) after addition of ascorbic acid (specimens 8, 10, 12, and 14) leads to high plasma levels of Cr with significantly lower Cr values in the erythrocytes than in samples without ascorbic acid (specimens 4 and 6). If ascorbic acid is added after spiking with Cr(VI) (specimens 16, 18, 20, and 22), the Cr concentrations in erythrocytes are comparable to those without ascorbic acid.

- After initial spiking with a high concentration of Cr(III) (500 μg), some Cr is found in the erythrocytes (specimens 7, 11, 15, 19, and 23), and the concentrations with and without addition of ascorbic acid are quite similar. (It is not certain that the Cr(III) was *in* the erythrocytes in these samples, it may have been only attached to the surface of the erythrocyte membrane).

TABLE 1. CHROMIUM CONCENTRATIONS IN WHOLE BLOOD, PLASMA AND
ERYTHROCYTES AFTER SPIKING OF BLOOD WITH CR(VI) OR CR(III) WITH
AND WITHOUT ASCORBIC ACID (AA) ADDITION

Specimen[a]		Substrate		Spiked to:		Results:		
		AA mg/l	Cr(VI) μg/l	Cr(III) μg/l	AA mg/l	μg Cr/l plasma[b]	μg Cr/l erythrocytes[b]	μg Cr/l blood
1	blank					< 1	< 1	< 1
2	AA	2.0				< 1	< 1	
3	AA	5.0				< 1	< 1	
4	Cr(VI)		50			32	17	50
5	Cr(III)			50		48	< 1	51
6	Cr(VI)		500			249	245	510
7	Cr(III)			500		455	36	520
8	AA+Cr(VI)	2.0	50			39	11	
9	AA+Cr(III)	2.0		50		48	< 1	
10	AA+Cr(VI)	2.0	500			262	206	
11	AA+Cr(III)	2.0		500		459	38	
12	AA+Cr(VI)	5.0	50			40	11	
13	AA+Cr(III)	5.0		50		50	< 1	
14	AA+Cr(VI)	5.0	500			293	207	
15	AA+Cr(III)	5.0		500		446	35	
16	Cr(VI)+AA		50		2.0	35	16	
17	Cr(III)+AA			50	2.0	47	<1	
18	Cr(VI)+AA		500		2.0	238	235	
19	Cr(III)+AA			500	2.0	455	29	
20	Cr(VI)+AA		50		5.0	35	15	
21	Cr(III)+AA			50	5.0	46	< 1	
22	Cr(VI)+AA		500		5.0	247	236	
23	Cr(III)+AA			500	5.0	433	39	

[a] EDTA has been added at 2.5 mg/l of blood sample.
[b] Adjusted for hematocrit.

As shown in Table 2, following 5 steps of dialysis, the Cr concentration in the erythrocytes of the blood samples highly spiked with Cr(III) is diminished to the level of unspiked control samples, whereas in the Cr(VI) spiked samples dialysis did not influence the Cr concentration in the red blood cells. This experiment demonstrates that:

• Cr(VI) enters the erythrocytes at a rate that is influenced by the plasma reductive capacity.

- Cr(VI) remains in the erythrocytes (presumably reduced and bound to protein structures, mainly to hemoglobin) even after several dialysis procedures.

- Cr(III) is detected in erythrocytes in only small amounts which are eluted completely by dialysis.

TABLE 2. CHROMIUM-CONCENTRATIONS IN ERYTHROCYTES (RBC) OF
SPIKED BLOOD SAMPLES FOLLOWING 1 TO 5 SEQUENTIAL DIALYSIS
STEPS WITH ISOTONIC SALINE

	Cr concentration in µg/l [a]		
Dialysing steps	Before spiking	After spiking to 500 µg/l blood	
		Cr(III)	Cr(VI)
	RBC	RBC	RBC
Pretreatment	10	-	-
1	8	45	198
2	7	20	190
3	6	12	192
4	5	6	196
5	6	6	206

[a] Adjusted for packed cell volume.

In other experiments (Table 3), we demonstrated that ascorbic acid concentration in the plasma is the predominant factor of the plasma reduction capacity, whereas glutathione plays a secondary role (Lewalter and Korallus, 1989).

Biomonitoring Results

We determined Cr levels in urine and whole blood in biomonitoring studies on workers with possible exposure to Cr (VI) during production and maintenance work in a plant manufacturing chromates and dichromates. We found two patterns of distribution: Group A with low blood and higher urine levels and group B with high blood and lower urine levels. Based on the *in vitro* studies described above, the determination of Cr in erythrocytes could be used as method of biomonitoring for internal Cr(VI) burden; therefore, we used this method for further biological monitoring. The methodology has been published elsewhere in detail (Lewalter et al., 1985).

TABLE 3. CHROMIUM COMPARTMENT DISTRIBUTION DEPENDENT ON EQUIMOLAR ASCORBIC ACID AND GLUTATHIONE ADDITION

Basic Blood Level		Blood Spiking			Cr-analysis *	
Cr*	µM AA/l	mM AA/l	mM GSH/l	Cr(VI)*	P	RBC
<1	15.7	-	-	5	4	<1
				50	20	32
				500	197	310
		10	-	5	5	<1
		10	-	50	47	2
		10	-	500	485	13
		-	10	5	3	2
		-	10	50	34	14
		-	10	500	380	121

* in µg Cr/l blood.

P=Plasma, RBC=erythrocytes
AA=ascorbic acid, GSH=glutathione

The average Cr(VI) airborne concentration in the facility was 0.045 mg/m^3. Where conditions of work were expected to lead to higher concentrations, respirators were worn. We found clear inter-individual differences in the relation between Cr concentrations in erythrocytes and the corresponding Cr concentrations in plasma and urine. Based on these inter-individual variations in plasma reductive capacity, we identified two groups of persons, the "strong reducers" and the "weak reducers," with results shown in the Table 4.

As predictable from *in vitro* studies, urinary excretion of Cr is higher for strong reducers than for weak reducers. In repeated analyses, we found that the designation of a person as either a "strong" or "weak" reducer is rather consistent. About two thirds of the workers we examined were considered to be strong Cr(VI) reducers, while approximately one third was rated as weak reducers.

TABLE 4. BIOMONITORING RESULTS OF STRONG AND WEAK REDUCERS
AT AVERAGE CR(VI) LEVELS OF 0.045 mg CR/m^3 AIR

Exposed persons	Number	Mean Cr concentrations		
		Plasma µg/l^a	Erythrocytes µg/l^a	Urine µg/l creatinine
All persons	76	5.2	6.8	9.1
Strong Cr(VI) reducers	52	7.3	2.5	12.2
Weak Cr(VI) reducers	24	4.0	11.2	5.9

^{a}Adjusted for hematocrit

Figure 2 presents data derived from our biological monitoring studies over recent years. It shows the relationship between Cr concentrations in urine, whole blood, plasma, and erythrocytes at different levels of airborne Cr(VI) for both strong and weak reducers. It is clear that determination of Cr in urine alone would lead to either an underestimate or an overestimate of the body burden due to occupational exposure to Cr(VI) compounds, depending on the plasma reduction capacity of the respective person.

DISCUSSION

Cr concentration in the erythrocytes is probably the best measure of an individual's relevant Cr(VI) burden. Cr(VI) penetration into the erythrocytes can occur at any time in the 3-month life span of the cells. Cr determination in urine merely presents a "snapshot" of the situation shortly prior to the time of measurement. When low Cr levels are found in the erythrocytes with correspondingly higher urinary Cr levels, extracellular reduction of the Cr(VI) can be considered to have been sufficient for rapid detoxification and elimination of these compounds.

For practical purposes of hazard evaluation, guidance values would be needed. The German MAK Committee has considered the development of Biological Tolerance Values (BAT) for hazardous substances. For alkaline chromates, correlations could be derived from existing data between air concentrations and Cr values in erythrocytes and urine under certain preconditions as shown in Table 5.

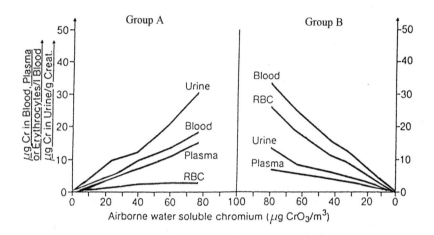

RBC=Erythrocytes

FIGURE 2. Correlation between airborne water soluble Cr(VI) concentrations and chromium concentration in blood, plasma, erythrocytes, and urine of workers in a dichromate plant at four different examinations differentiated according to the reduction capacity.

Group A="strong" reducers; Group B="weak" reducers.

For the purpose of hazard evaluation, it is useful to establish "normal" values for persons not occupationally exposed to Cr(VI) compounds. In unexposed persons, Cr concentration in erythrocytes up to about 1 μg/l could be detected. Therefore, values above that level would suggest an exposure to Cr(VI) during weeks prior to the sampling date. The association of a hazard may be estimated using the air limit value (TRK) assigned to the respective compounds using the equivalence table, whereby findings below the Cr concentration in the erythrocytes correlated to the TRK (0.10 mg/m^3) would be indicative of exposure without quantifiable hazard. One fourth of the TRK could be used as an action level to inspect the workplace and control the industrial hygiene procedures in order to confirm that all preventive measures are in place or to indicate the need for changes.

TABLE 5. EXPOSURE EQUIVALENTS FOR ALKALINE CHROMATES (VI)

Air CrO$_3$ (mg/m^3)	Chromium in Erythrocytes[a] (μg/l whole blood)	Chromium in Urine[b] (μg/l)
0.03	9	12
0.05	17	20
0.08	25	30
0.10	35	40

[a] Sampling time: long-term exposure, after several shifts
[b] Sampling time: end of exposure or end of shift

According to DFG (1993).

CONCLUSIONS

Our studies demonstrate the importance of measuring Cr levels in erythrocytes for an accurate assessment of an individual's internal Cr burden. After penetration of Cr(VI) into the erythrocytes, the resulting Cr(III) remains within the cells and cannot be eliminated even by repeated dialysis. This suggests that Cr compounds taken up in hexavalent form are stored within red blood cells for their lifespan. "Strong" and "weak" reducers could be identified in response to exposure to Cr(VI) compounds. Exposure to Cr(VI) compounds can be monitored by Cr determination in erythrocytes in parallel to the determination of Cr in urine as a method to detect recent Cr exposure.

These findings apply mainly to highly water-soluble chromates assumed to have a relative short retention time in the alveolar cells. It is not clear whether these results are also applicable to water-insoluble

Cr(VI) compounds. It may be assumed that the Cr content in erythrocytes representatively describes the Cr burden in other target cells, although this has not been validated experimentally.

REFERENCES

Cohen, M. D., B. Kargacin, C. B. Klein, and M. Costa. 1993. Mechanisms of chromium carcinogenicity and toxicity. Critical Reviews in Toxicology 23 (3):255-281.

DFG. 1993. Maximum concentrations at the workplace and biological tolerance values for working materials. VCH Weinheim 29:95.

Korallus, U., C. Harzdorf, and J. Lewalter. 1984. Experimental basis for ascorbic acid therapy of poisoning by hexavalent chromium compounds. International Archives of Occupational and Environmental Health 52:247-256.

Lewalter, J., U. Korallus, C. Harzdorf, and H. Weidemann. 1985. Chromium bond detection in isolated erythrocytes: a new principle of biological monitoring of exposure to hexavalent chromium. International Archives of Occupational and Environmental Health 55:305-318.

Lewalter, J., U. Korallus, C. Harzdorf, and H. Weidemann. 1986. Möglichkeiten der Metallspurenbestimmung zur Bewertung innerer Metallbeanspruchungen. Band 2, VCH Verlagsgesellschaft mbH, Weinheim.

Lewalter, J., and U. Korallus. 1989. The significance of ascorbic acid and glutathione for chromate metabolism in man. Toxicology and Environmental Chemistry 24:25-33.

Contributors

Richard Albertini, M.D., Vermont Cancer Center, UVM, 32 No. Prospect Street, Burlington, VT 05401, USA.

Paul Brandt-Rauf, M.D., Sc.D., Dr.P.H., Division of Environmental Sciences, Columbia University, 60 Haven Ave., New York, NY 10032, USA.

Charles R. Cantor, Ph.D., Center for Advanced Research in Biotechnology, Boston University, 36 Cummington St., Boston, MA 02215, USA.

Ken Chadwick, Ph.D., Directorate General for Science, Research and Development, Commission of the European Communities, Rue de la Loi 200, B-1049 Brussels, Belgium.

Morton Corn, Ph.D., Department of Environmental Health Engineering, Johns Hopkins School of Hygiene and Public Health, 615 North Wolfe Street, Baltimore, MD 21205, USA.

Immaculata DeVivo, Ph.D., Division of Environmental Sciences, Columbia University, 60 Haven Ave., New York, NY 10032, USA.

Joe Gray, Ph.D., Department of Laboratory Medicine, University of California/San Francisco, San Francisco, CA 94143-0808, USA.

Frank Gonzalez, Ph.D., National Cancer Institute - NIH, 6130 Executive Blvd., Rockville, MD 20852, USA.

Philip Harber, M.D., M.P.H., Occupational and Environmental Medicine, UCLA, 10911 Weyburn Ave. #344, Los Angeles, CA 90024, USA.

Thomas W. Henn, M.D., M.P.H., Hanford Environmental Health Foundation, 3080 George Washington Way, Richland, WA 99352 USA.

Ronald Jensen, Ph.D., Life Sciences Division, Lawrence Berkeley Laboratory, Berkeley, CA, USA.

James H. Jett, Ph.D., National Flow Cytometry Resource, Mail Stop M-888, Los Alamos National Laboratory, Los Alamos, NM 87545, USA.

Karl Kelsey, Ph.D., Harvard University, School of Public Health, 665 Huntington Avenue, Boston, MA 02115, USA.

Marvin S. Legator, Ph.D., Department of Preventive Medicine and Community Health, University of Texas Medical Branch at Galveston, Galveston, TX 77555-1010, USA.

J. Lewalter, Ph.D., Leitender Werksarzt der Bayer AG, Bayer AG, WV-LE Arztliche Abteilung, D-5090 Leverkusen, Bayerwerk, Germany.

Howard L. Liber, Ph.D., Department of Cancer Biology, Harvard School of Public Health, Boston, MA 02115, USA.

Joe N. Lucas, Ph.D., Lawrence Livermore National Laboratory, Livermore, CA 94550, USA.

Lewis Maltby, J.D., National Task Force on Civil Liberties in the Workforce, American Civil Liberties Union, 132 W. 43rd. Street, New York, NY 10036, USA.

Marie-Jeanne Marion, Ph.D., Unité de Recherche sur les Hépatites, le Sida et les Retrovirus Humains, INSERM, Lyon, France.

Robert J. McCunney, M.D., M.P.H., Environmental Medicine Services, Massachusetts Institute of Technology, 18 Vassar Street, Cambridge, MA 02139, USA.

Mortimer Mendelsohn, M.D., Ph.D., Radiation Effects Research Foundation (RERF), 5-2 Hijiyama Park, Minami-ku, Hiroshima 732, Japan.

Leopold W. Miksche, Dr. Med., Leitender Werksarzt der Bayer AG, Bayer AG, WV-LE Arztliche Abteilung, D-5090 Leverkusen, Bayerwerk, Germany.

Dan Moore, Ph.D., California Pacific Medical Center, San Francisco, CA, USA.

Ken Olden, Ph.D., Director, NIEHS (B1-02), P.O. Box 12233, Research Triangle Park, NC 27709, USA.

Frederica Perera, Dr. P.H., School of Public Health, Columbia University Comprehensive Cancer Center, 60 Haven Avenue, B-1, New York, NY 10032, USA.

James Piper, Ph.D., Medical Research Council, Edinburgh, Scotland, UK.

Charles S. Rabkin, M.D., Viral Epidemiology Branch, National Institutes of Health, 6130 Executive Blvd., Rockville, MD 20852, USA.

Venkateswara Rao, Ph.D., Science Applications International Corporation, 7600-A Leesburg Pike, Falls Church, VA 22043, USA.

William Rom, M.D., Pulmonary Division, Bellvue Hospital Center, 27th Street & 1st Avenue, New York, New York 10016, USA.

Milton Rossman, M.D., Hospital of the University of Pennsylvania, 3400 Spruce Street, Philadelphia, Pennsylvania 19104, USA.

Nathaniel Rothman, M.D., National Cancer Institute, Occupational Studies Section, 6130 Executive Blvd., Rockville, MD 20892, USA.

Mark Rothstein, J.D., Health Law and Policy Institute, University of Houston Law Center, 4800 Calhoune, Houston, Texas 77204-6381, USA.

Cesare Saltini, M.D., University of Modena, Institute of Tuberculosis and Respiratory Diseases, Via Del Pozzo, 71, 41100 Modena, Italy.

Takeshi Sano, Ph.D., Center for Advanced Biotechnology, Boston University, 36 Cummington Street, Boston, MA 02215, USA.

Paul Schulte, Ph.D., National Institute for Occupational Safety and Health, 4676 Columbia Parkway, Cincinnati, OH 45226-1998, USA.

Jaak Sinnaeve, Ph.D., Directorate General for Science, Research and Development, Commission of the European Communities, Rue de la Loi 200, B-1049 Brussels, Belgium.

Cassandra L. Smith, Ph.D., Center for Advanced Biotechnology, Boston University, 36 Cummington Street, Boston, MA 02215, USA.

Walter F. Stewart, Ph.D., Department of Epidemiology, Johns Hopkins University School of Hygiene and Public Health, Baltimore, MD 21205, USA.

Tore Straume, Ph.D., Health and Ecological Assessment Division, Lawrence Livermore National Laboratory, Livermore, CA 94550, USA.

Vera Thomas, Ph.D., University of Miami School of Medicine, Department of Anesthesiology, P.O. Box 016370, Miami, FL 33101, USA.

James E. Trosko, Ph.D., Department of Pediatrics and Human Development, Michigan State University, East Lansing, MI 48824, USA.

Alan Unger, Ph.D., Science Applications International Corporation, 7600-A Leesburg Pike, Falls Church, VA 22043, USA.

John K. Wiencke, Ph.D., Department of Epidemiology and Biostatistics, University of California at San Francisco, San Francisco, CA, USA.

ABBREVIATIONS

ACGIH	American Conference of Governmental Industrial Hygienists
ADA	Americans with Disabilities Act
AFB	Aflatoxin B
ANLL	Acute non-lymphocytic leukemia
ASL	Angiosarcoma of the liver
ATSDR	Agency for Toxic Substances and Disease Registry (USA)
BAT	Biologische Arbeitsstofftoleranzwerte
BEI	Biological exposure indices
c-AMP	Cyclic adenosine monophosphate
CAA	Chloroacetaldehyde
CBD	Chronic beryllium disease
CDC	Centers for Disease Control and Prevention (USA)
CEC	Commission of the European Communities
CEO	Chloroethylene oxide
CGH	Comparative genomic hybridization
CI	Confidence interval
CYP	Cytochrome P
DNA	Deoxyribonucleic acid
DOD	Department of Defense (USA)
DOE	Department of Energy (USA)
EEOC	Equal Employment Opportunity Commission (USA)
ELISA	Enzyme-linked immunosorbent assay
EMF	Electromagnetic field
EPA	Environmental Protection Agency (USA)
EROD	Ethoxyresorufin-o-deethylase
FDA	Food and Drug Administration (USA)
FISH	Fluorescence in situ hybridization
GJIC	Gap junctional inter-communication
GPA	Glycophorin A
GST	Glutathione S-transferase
GTP	Guanosine triphosphate

HASP	Health and safety program
HCC	Hepatocellular carcinoma
HEG	Homogeneous exposure group
HIV	Human immunodeficiency virus
HLA	Human leukocyte antigens
HPLC	High performance liquid chromatography
HPRT	Hypoxanthine phosphoribosyl transferase
IC	Intercellular communication
LDH	Lactate dehydrogenase
LET	Linear energy transfer
LPT	Lymphocyte proliferation test
LTT	Lymphocyte transformation test
MHC	Major histocompatibility complex
MNU	Methyl-nitrosourea
MTBE	Methyl tertiary butyl ether
NAS-NRC	National Academy of Sciences-National Research Council
NASA	National Aeronautics and Space Administration (USA)
NAT	N-Acetyl transferase
NIEHS	National Institute for Environmental Health Sciences (USA)
NIH	National Institutes of Health (USA)
NIOSH	National Institute for Occupational Safety and Health (USA)
NRC	National Research Council (USA)
NTP	National toxicology program (USA)
OCDD	Octachlorodibenzo-p-dioxin
OODB	Object-oriented database
OSHA	Occupational Safety and Health Administration (USA)
OTA	Office of Technology Assessment (US Congress)
PAH	Polycyclic aromatic hydrocarbons
PCB	Polychlorinated biphenyls
PCDD	Polychlorinated dibenzo-p-dioxin
PCR	Polymerase chain reaction
PHA	Phytohemagglutinin
PPE	Personal protective equipment
PRC	Plasma reduction capacity
RAST	Radioallergosorbent testing

RERF	Radiation Effects Research Foundation (Japan)
RNA	Ribonucleic acid
RR	Relative risk
SCE	Sister chromatid exchange
SSOP	Sequence specific oligonucleotide probe
TCDD	Tetrachlorodibenzo-p-dioxin
TCDF	Tetrachlorodibenzofuran
TCR	T cell receptor
TEF	Toxicity equivalence factor
TLV	Threshold limit value
TWA	Time-weighted average
VC	Vinyl chloride
WCP	Whole chromosome probe

Index